CRC Mathematical Modelling Series

Series Editor
Nicola Bellomo
Politecnico di Torino, Italy

Advisory Editorial Board

Titles included in the series:

Designing Innovations
in
Industrial
Logistics
Modelling

Edited by

Andrew Kusiak
Maurizio Bielli

CRC Press

Boca Raton   New York   London   Tokyo

Acquiring Editor: Tim Pletscher
Senior Project Editor: Susan Fox
Marketing Manager: Susie Carlisle
Direct Marketing Manager: Becky McEldowney
Cover design: Dawn Boyd

Library of Congress Cataloging-in-Publication Data

Designing innovations in industrial logistics modelling / edited by
   Andrew Kusiak and Maurizio Bielli
        p.   cm. — (CRC mathematical modelling series)
      Includes bibliographical references and index.
      ISBN 0-8493-8335-8 (alk. paper)
      1. Production control—Mathematical models.   2. Materials
   management—Mathematical models.   3. Business logistics—
   Mathematical models   I. Kusiak, Andrew.   II. Bielli, Maurizio
   III. Series.
   TS157.D47 1996
   658.5′01′5188—dc20   96-34581
                                                                      CIP

© 1997 by CRC Press, Inc.

No claim to original U.S. Government works
International Standard Book Number 0-8493-8335-8
Library of Congress Card Number 96-34581
Printed in the United States of America 1 2 3 4 5 6 7 8 9 0
Printed on acid-free paper

Preface

Industrial logistics represents an emerging discipline of increasing interest in the manufacturing environment, thanks to its capabilities of integrating different methodological approaches and techniques in order to design and manage innovations at structural, technological, and organizational levels. In fact, industrial logistics services are becoming recognized as an essential element of customer and company satisfaction in a growing number of product markets today, and are one of the strategic suprasystems that are responsible for creating customer value and sustainable competitive advantages. Moreover, industrial logistics quality and integrated logistics system management represent other relevant topics where models, tools, and computation techniques, such as benchmarking, optimization, and simulation, should play a preeminent role in the quality process analysis. In fact, it is only through understanding the tools at their disposal that logistic managers can make intelligent analysis decisions.

In this overall context, this book collects a selection of contributions presented at the CIRP-IFORS International Workshop on "Designing Innovations in Industrial Logistics: Models and Methods", held in Capri (Italy) on 29 September - 1 October 1993.

Its contents are organized into four chapters collecting papers from university research centers and industrial companies, and dealing with problems, methods, models, case studies, and projects developed in the manufacturing, and logistic settings.

The first chapter deals with problems and trends in designing new logistic systems and presents models and approaches to manage and evaluate the dynamic evolution and the structural changes of production systems.

The introductory paper by Villa and Bielli gives an overview of the research perspectives in industrial logistics design. A new formulation of the organization of design subproblems is outlined, that is the layout and management strategy design. Then, a discrete-event process based on marginal innovation introduction is considered and a mathematical model of the design process planning is illustrated, and the interactions with concurrent engineering procedures are discussed.

The following paper by Luhtala and Eloranta addresses typical problems of managing order-bound — that is converging instead of diverging — supply chains, and suggests a time-based approach as the focus for development actions. The relative benefits are studied especially in the transition phase, when the supply chain is streamlined from a group of physical factories into a global logical factory. Then, an analogy between a factory and an order-bound supply chain is described and a concept of time profit is introduced as an instrument for matching the structure and the control practices of the supply chain during the transition phase.

The third paper by Lucertini and Telmon analyzes a method to represent production systems based on three entities: the process (what the system is asked to do and which resources are needed) represented as a combination of two sets of flows, the organization (how the resources are managed), the decision process (who does the job, when and how) that drives the behavior of the production system. The paper makes an attempt to outline new paradigms for representing strategic organizational structural changes and for defining performance indicators which take into account in an integrated way the three entities.

The second chapter concerns the methods and tools suitable in designing and managing innovation programs in industrial logistics.

The problem of setting decision rules for experts reallocation has been tackled in the paper by Belhe and Kusiak. In fact, in the design team a particular expertise may be required for a certain time, to make a contribution towards achieving a design goal. Due to the high cost of certain expertise, the expert's inclusion may not be justified at all the stages of product development. The activity status is checked periodically to assess the requirement of the expert's services for the next time period. Two different approaches to the decision-making process are presented. In the first approach, the process of managing expert resources is modeled as a Markov decision process over a finite horizon, while the second approach is based on the optimal stopping of Markov chain. The optimality criteria are based on previous theoretical developmets.

Then, Petri-net models for simulation and analysis of complex logistic systems and business process management are illustrated in the paper by Wagner and Bullinger. Their advantages in dealing with discrete state and time systems are discussed and an application to modeling and planning flexible assembly systems is presented.

Moreover, a scenario modeling approach to solve stochastic production planning problems is presented in the paper by Escudero and Kamesan. Linear programming models and a methodology to solve them are presented. Uncertainty in demand is used as a test case and is characterized via scenarios; then, individual scenario solutions are aggregated to yield an implementable non-anticipative policy. Such an approach makes it possible to model correlated and non-stationary demand as well as a variety of recourse decision types. For computational purposes, two alternative representations are proposed: a compact approach suitable for the Symplex Method and a splitting variable approach suitable for the Interior Point Method. A crash procedure is developed and computational results are reported and discussed.

The third chapter presents some case studies carried out by companies and universities involved in research projects at national and European levels.

In the paper by Calderini, Cantamessa, and Nicolò the problem of designing Tool Logistic Systems in flexible manufacturing environments is addressed and a methodological approach is illustrated in detail with the description of attributes and modules.

Moreover, in the paper by Ukovich, Crasnich, and Zanetti, an R&D project to design a logistic integrated network for decision enhancement (called LINDEN) is presented. Its aim is to provide a decision support tool for the logistic managers in each point of the network for all decision levels: strategic, tactical, and operational. The characteristics of LINDEN operational aspects are outlined and the project perspectives discussed.

Then, an integrated approach to support decision-making in designing innovations in industrial logistics is presented in the paper by Crasnich, Lanza, and Merli. The logistic workbench approach and some specific modules are illustrated and discussed, with a special emphasis devoted to the frequency module, characterizing the mathematical programming models for transportation logistics and vehicle routing.

In the last paper of this chapter by Rolstadas and Strandhagen, a conceptual model of design choices, intermediate variables, and performance indicators is presented, particularly useful to design or test a one of a kind and batch production system, based on a generalization of Walras economical model, identifying basic data structures, products, and resources.

The fourth chapter of the book concerns the introduction of advanced information technologies to manage and promote innovations in industrial logistics.

The paper by Tse defines the process innovation as a drastic change in the links within the order-process-distribution chain that allows the company to change the rules of the game to its favor in the global dynamic competitive environment. Process innovation involves activities which are information intensive, and advanced information technology, if properly managed, should be an important enabling factor to promote process innovation. This contribution develops a dynamic model for process innovation, describing the types of information activities in searching for innovation potential, finding innovative solutions and scaling them up to spread the effect of innovation. Then, how information technology should be directed and managed to promote process innovation is discussed.

Finally, the paper by Bansal deals with innovation management in CIM environments, where innovative and supportive management functions have to impact their own immediate upstream process and also the ultimate driver, i.e., the dynamic customer demand, with the value they can add.

Justification and implementation paradigms of innovation management in these technologies are explored, on the basis of actual project experience. Research issues are also elaborated and articulated.

The Editors express their gratitude to the authors of this volume. Without their contribution the book would not have been possible.

M. Bielli and A. Kusiak

Contents

**Part 4 - Information technologies for managing and
promoting innovations in logistics**

PART 1

PROBLEMS AND TRENDS IN THE DESIGN
OF INNOVATION PROGRAMS IN LOGISTICS

Chapter 1

RESEARCH PERSPECTIVES
IN INDUSTRIAL LOGISTICS DESIGN

Agostino Villa
Dipartimento di Sistemi di Produzione
ed Economia dell'Azienda - Politecnico di Torino, Italy

Maurizio Bielli
Istituto di Analisi dei Sistemi ed Informatica
Consiglio Nazionale Ricerche, Roma, Italy

1. INTRODUCTION

The possibility of utilizing a good logistic system in an industrial plant appears to be a key factor for the successful implementation and management of a flexible manufacturing plant. Owing to the crucial importance of flexibility when turbulent market demands have to be satisfied, high capability of the logistic system to servicing "just in time" the requests for tools and components of the machining centers is an absolute need.

Often in the past the design of an *Industrial Logistic System* (*ILS*) has been considered a secondary task to be solved once the arrangement of the machining centers and of the storage places in the department space has been completed. Recently, industrial managers have realized that the amplitude of demand variabilities directly reflects into a wider product mix and then into higher requests for tool and part movements [1,2], thus making the ILS design a crucial task for assuring a successful manufacturing process.

Attempts made for reducing the time spent at machining centers in waiting for tools and components are showing that the problem of obtaining a "good" ILS cannot be solved at all by only designing a good network of connections among machining centers but it also requires defining a sound strategy for managing movements of parts and tools in time [3]. Moreover it has been recognized in practice (and it appears to be quite obvious also from a theoretical viewpoint) that only a combined optimized design of the ILS structure (i.e., layout) and of a specific ILS management strategy can assure the industrial end user an *Efficient-Effective-Economic (E3)* logistic service [4].

From both practical and theoretical points of view, the joint design of an

ILS layout and the related ILS management system appears to be a very complex task involving several design subproblems:

(A)   *ILS layout design subproblems*, i.e.
   • drafting the ILS network of connections;
   • planning the average production flows;
   • planning the buffers sizes;
(B)   *ILS management design subproblems*, i.e.
   • designing a master scheduling strategy;
   • designing a distributed service and dispatching control strategy.

The above outlined decomposition has been mainly originated by practitioners and managers operating in industrial shop-floors: as a consequence, it suffers the intrinsic weakness of heuristic approaches.

Now theoreticians are requested to formulate prescriptive models of the complete set of design subproblems and of their interrelations (we call this set an *organization of design subproblems*) in such a way that the individual design strategies (obtained by solving each subproblem individually) as well as the convergence of all the organization of said design strategies can be justified.

The goal of this chapter is to outline a new formulation of the mentioned *organization of design subproblems* and to focuse on the most crucial subproblems which industrial managers need to solve by applying robust and guaranteed solution procedures.

Main lines of the chapter are as follows.

In principle, an *organization of design subproblems* can be modeled as a table of relations among the subproblems themselves such that each entry of the table contains a model of the data and constraints which one subproblem induces on another one. A similar table describes relations among the solution strategies of the above mentioned subproblems, showing for each strategy the design resources required for its implementation.

This concept of subproblems/strategies tables focuses on two main industrial tasks to be tackled in the ILS design, namely:
   • managing the resources required for developing and implementing an organization of design strategies;
   • developing effective design strategies for the most critical individual subproblems.

When considering each one of the two tasks in an industrial frame, a proper line of evolution can be recognized. The former evolves towards utilization of *Concurrent Engineering* tools [5]. The latter shows a growing interest of manufacturing plant managers in *modular organization of the manufacturing plant* [6].

Which may be the most significant motivations of these evolution lines?

This chapter tries to propose answers and suggests research perspectives.

2. ILS DESIGN: LOGICAL STATEMENT

In dealing with the ILS design task, industrial managers' requirements imply that two main problems have to be approached:

(A)   given a set of requirements specifying the expected service target and a technological data base including information on the available ILS components, it's required to *design the ILS layout* by detailing the connections, routes, and storage places, as well as the types of devices to be used for moving and storing parts and tools in the manufacturing department;

(B)   assumed to have at disposal an ILS layout and given specific scenarios for the parts and tools supply demands, it's required to *design the ILS management strategy* by detailing how parts and tools have to be addressed on alternative routes and stored in available buffers in such a way to avoid blocking and starving situations at machining centers and assuring their maximum possible utilization for given release time and due date constraints.

The two problems are strictly connected together because the design of an ILS layout cannot be completed and validated except in the case of an ILS management strategy that has been adapted on the selected layout. On the other hand, an ILS management strategy cannot be effectively optimized except in cases where an appropriate ILS layout could be adapted to the demands scenario (Figure 1).

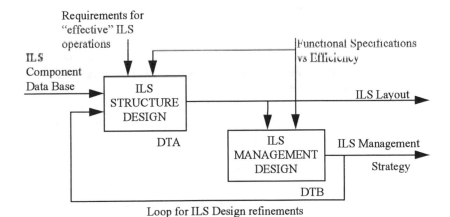

Figure 1. Interactions among the phases of ILS Layout Design and ILS Management Strategy Design.

This second consideration appears to be particularly true when a net-based ILS layout has to be designed and optimally managed for a wide variety of parts and tools to be moved and fast variations of the part/tool demands.

According to a more detailed view, the ILS layout design can be organized into three steps:

(A-1)   organization of the network of connections, that means definition of the graph of admissible connections among the machining centers and their services (as tool room, buffers and QC/measurement services), based on the technological constraints specified by the process plans to be implemented;

(A-2)   preliminary evaluation of the average loading conditions for the desired ILS, assuming a given scenario of the part/tool demands, and using the simplest possible model of the part/tool movements in the manufacturing department, and consequent planning of the ILS storage and transportation capacities in such a way to assure that the service requirements could be satisfied with the maximum possible efficiency;

(A-3)   organization of the ILS layout by verifying the consistency between the storage and transport capacities required and the capabilities of available technologies and components; more precisely ILS layout organization implies:

- verification of the opportunity of utilizing all possible locations and types of storages proposed in step (A-1);

- evaluation of the convenience of utilizing different devices on same links for transporting different items (e.g. parts and tools), and then synchronization of the supplies of the required items at each machining node;

- or, on the contrary, evaluation of the convenience of utilizing same devices on the same links;

- comparison of the obtained layout with the ones already at disposal of the designer (and stored in a suitable technological data base), in such a way that the adopted solution could be either confirmed by its commonality with previous ones, or be denied for final decision to a better detailing, when innovative.

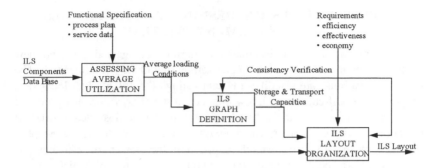

Figure 2. Details of the ILS Structure Design task.

The ILS management design phase mainly refers to the problem of scheduling the service activities for a given ILS layout.

In principle, this design task is clearly stated, while in practice an industrial-oriented approach to this task is very complicated because any ILS management strategy should concurrently handle two different types of production flows: parts and tools at each work unit.

In fact, a work unit is intended to be a service center at which manufacturing operations must synchronize the supplies of the two mentioned different types of items by managing the two respective routes.

Synchronization of part and tool supplies at each machining unit in a shop floor is a very hard task because of both the complexity of interactions among the different units and the difficulty of coordinating the various supply events, both at the planning and at the control level [7]. By this reason the ILS management design is usually decomposed into several steps, ranging from mid-term scheduling of production orders, short-term dynamic scheduling of small batches in the shop floor, up to Production Activity Control-PAC [8].

According to the outline of the ILS design task sketched above (Figure 2), one can easily realize that the two crucial design problems mentioned at the end of Section 1 really appear to be key steps for a successful design. First of all, availability of a methodology to model and optimize the management system for an ILS, possibly arranged as a modular organization of intelligent service units (i.e., machining, storing, dispatching, and testing units) could be a bridge toward effective development and utilization of reactive scheduling procedures. Second, a methodology to optimize the management of a complex ILS design, represented as an organization of several design activities (i.e., a *design process*) requiring many different resources (as the above proposed logical statement suggests), seems to be the real necessary tool for an efficient and effective approach to the ILS design task.

The two following sections will better focus on these two concepts.

3. PERSPECTIVES ON DESIGN OF MODULAR ILS MANAGEMENT SYSTEMS

In several industrial environments it is now recognized that the new frontier in manufacturing plant organization should be "modularity". The main motivations are facility in innovating the plant by local and progressive modifications with low impacts on the shop floor, and potential simplicity in managing the network of intelligent modules, clearly oriented to cooperate with each other. The main obstacle to the modularity concept is the lack of models and methods supporting the design of efficient and effective management strategies for a plant the structure of which can be modified in time either due to failures or innovative programs. Practitioners expect that only in the presence of a network of controllers, as in the case of a controller for each machining unit with a strong cooperation with each other, can effective management be assured.

Said ideas have inspired research in different complementary fields such as the ones in *cooperative systems* [9], *hierarchical control structures* [10], *object-oriented programming* [11], *real-time negotiation of resource utilization* [12], and *opportunistic scheduling* [13].

In the authors' opinion, these research lines should be greatly improved.

This suggestion can be easily justified by the following brief outline of the ILS management system design problem.

Assume that the goal of the ILS management system design is to make a new manufacturing system composed of groups of machines working together in a coordinated way in order to reach a common objective (as production throughput), and also organized in such a way to be easily modified through embedding of new machines. Moreover suppose that:

- the manufacturing system task is to accomplish a number of jobs according to given delivery constraints;
- for each job a sequence of precedence constraints among the operations to be performed at the different machines has been stated by a Computer Aided Process Planning - CAPP procedure, thus defining for each job the sequence of machines by which it will be processed.

For each machine, with reference to each job assigned to itself, it is assumed is the following data known:

- the list of operations to be accomplished;
- the preceding operation (and the assigned upstream machine) and the subsequent operation (and assigned downstream machine) together with the expected time of job delivery from upstream and expected due date for job release at downstream;

- penalty weights in case jobs are released too late to downstream and in case of jobs scheduled too early and not yet received from upstream.

In principle, on the basis of this information, each local controller of a machine can autonomously decide the sequence of operations to be performed in order to optimize a local target defined according to penalty weights mentioned above. In addition, each local controller has to communicate its own locally optimized times of delivering parts to downstream as well as its own desired times of receiving parts from upstream, in such a way that neighboring controllers could optimize their service strategies accordingly.

Because of the weights nature and the information pattern structure (it looks to be a *complete information pattern* [14]), this organization of modular local strategies accounts for the overall goal of the ILS management. Then it has to be proved that the set of locally optimized strategies corresponds to an accurate suboptimal solution of the overall ILS management problem.

The network of modular controllers should be characterized by the following properties:

- the local knowledge based on information from neighboring machines and the parametric knowledge of the overall system goal is sufficient for a short-term near-optimal selection of the sequence of operations to be performed by the machines;
- the dynamic refreshing of this selection can bring to a near-optimal behavior of the overall system.

According to this conjecture, one can expect to also obtain the following consequences:

- there is no need for a centralized scheduler but just for a central CAPP;
- the self-organizing set of local controllers is robust against potential disturbances and modifications of individual machines, thus allowing reconfigurations of the plant through progressive modifications applied locally.

To prove above conjectures and expected results, we suggest that the organization of modular controllers has to be designed according to a *contract net* model [15] which seems to be the most efficient approach in distributed control of discrete event dynamic systems, as manufacturing plants are.

4. PERSPECTIVES ON MANAGEMENT OF COMPLEX ILS DESIGN PROCESS

The second key problem to be solved for efficiently approaching the design of a complex ILS is to develop a robust methodology for the management of all the activities and resources involved in the design process.

Evidently, this problem can be generalized to any complex system design. Nevertheless it has received till now insufficient attention. Often only heuristics has been used to formulate design management strategies without any generality and suboptimality property. But now experience-based design appears to be insufficient if not supported by formal tools, and this consideration is particularly true in case of ILS design owing to the large number of design activities and resources there involved. A brief summary of the most evident ILS design phases can justify this assertion.

In practical terms, designing an "effective" ILS implies that:

- ILS must be provided with sufficient stock of tools of the required types;
- ILS transportation capacity must be sufficiently dimensioned;
- ILS monitoring capability must be sufficiently accurate;
- ILS supply operations must be synchronized with the production management timing.

In formal terms, designing an "effective" ILS implies that a complete design of a discrete-event dynamic system will be made, where "complete design" means that both the ILS layout and the ILS management strategy must be jointly selected and validated.

In theoretical terms, designing an "effective" ILS means to solve a combinatorial search for the ILS structure components, conditioned on the hypothesis of a given production plan (then of given demands for tools) together with a scheduling problem for the demanded parts and tools.

The above outlined three different types of approaches (i.e., "formal", "practical", and "theoretical") suggest *to base the ILS design on the search for a good trade-off among the following ILS features*:

- enhanced transportation capacity, but at a reasonable cost;
- sufficient storage capacity, but with low volumes, space, and cost;
- reduced tooling costs;
- efficient overall manufacturing system performance.

Therefore, the ILS design process, basically consisting of a multiobjective search process, can be transformed into a set of interconnected optimization subproblems by selecting for each subproblem a specific objective and transforming the others into constraints. The subproblem solutions are then iterated until a satisfactory compromise is obtained.

This makes it useful to approach the following three ILS design phases (see Figure 3):

1. ILS structure generation;
2. ILS management strategy optimization;
3. ILS efficiency estimation.

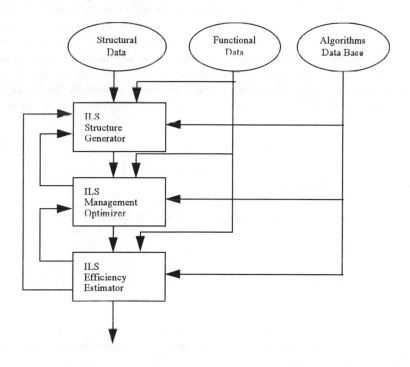

Figure 3. Organization of the ILS design phases.

The purpose of the ILS structure generation phase is to select the layout and the physical components of the part and tool transportation and storage network, by analyzing all possible and convenient connection patterns among the machining units together with the tool room, in order to assure the required efficiency within the assumed production scenario.

Both exogenous (i.e., provided by the designer) and endogenous (i.e., coming from the execution of other design phases) inputs are required. Exogenous inputs supply data describing admissible layouts and physical components, data representing the part programs, technological characteristics of tools to be handled, and production requirements.

Endogenous inputs contain suggestions for improving the ILS structure design, coming from phase 3, and suggestions for improving the ILS structure parameters (e.g., transportation and storage capacities) coming from phase 2.

Outputs of the ILS structure generation are a first ranking of ILS layout proposals, in terms of feasible transportation and storage networks, as well as time-dependent production scenarios, and in terms of production order schedules which execution can be assured by tool supplies made by the proposed ILS networks.

The purpose of the ILS management strategy optimization is to optimize the utilization of the transportation and storage network, proposed in phase 1, by selecting scheduling and control strategies among a library of models and algorithms. Exogenous inputs of this phase are the ILS performance specifications, to be used as benchmarks in the validation of the tool management strategy. Endogenous inputs are the ILS structure (from phase 1) and suggestions for modifications of the ILS management strategy (from phase 3).

Output of this phase include a second ranking of ILS proposals, joint schedules of tools and parts, and suggestions for improving the ILS structure, to be sent to phase 1 in case the optimized management strategy cannot reach desired specifications.

The purpose of the ILS efficiency estimation phase is to perform the most accurate ranking of ILS proposals to be submitted to the designer for final acceptance. To this aim, admissible ILSs are evaluated according to the whole set of relevant efficiency-effectiveness-economy criteria. Exogenous inputs of this third design phase are the performance specifications to be used as benchmarks for the ILS selection. Endogenous inputs are both the second ranking of ILS proposals and the detailed tool-part schedules (coming from phase 2).

Outputs of phase 3 are the final ranking of ILSs, both in terms of structures and management strategies, and the suggested modifications, thus specifying all the attributes of the desired ILS.

Let us simply refer in the following to the ILS design phase.

In general, designing an ILS layout means to compute a set of *attributes* which completely specify the ILS physical organization.

Assume that the computation of each attribute k can be performed by applying a specific *design module* and by using some *design resources R(.)* among the available ones.

Then, the design process consists of a discrete-event process in which each design module, applied at a discrete time, gradually modifies the *attribute state*.

As soon as a target value will be reached by an attribute state, pushed by the innovation action applied by a selected design module, then the desired attribute will be obtained and a *marginal innovation* of the ILS will be accomplished.

In general, more than one attribute state can be simultaneously subject to modification, by applying more than one design module. In this case, a sharing of the design resources among the different design modules is necessary.

Each time either an attribute state reaches its target value or a new design module is activated, a different utilization of the design resources has to be planned. As a consequence, the design process appears to be a sequence of events at which a modification of the design resources utilization has to be managed. The time between two consecutive events can be denoted *design session*.

Let us introduce the following notations:

$k=1,...,NA$, design attribute index, corresponding to a related design module;
$p=1,..., 2NA$, design session index;
$X(k,p)$, design attribute state, when starting session p;
$S(p)$, time instant at which session p starts;
$R(k,p)$, amount of design resources dedicated to the design module k during session p;
$A(k)$, initiation time when using the design module k;
$Z(k)$, termination time in using module k;
ST, start-up time of the design program;
DD, due date for final completion of the design program;
R, total amount of available resources.

According to these notations, *planning the design program means to decide*:

- for each design module k, when its utilization must start, i.e., $A(k)$, $k=1,...,NA$;
- for each design session p and each design module k, expected to be used in said session p, how large the amount of required design resources must be, i.e., $R(k,p)$, for any k and p.

Effectiveness of the design program implies satisfying the following conditions:

$$\min_{k}\{A(k)\} \geq ST \tag{1}$$

$$\max_{k}\{Z(k)\} \leq DD \tag{2}$$

The design process can be considered a sequence of design sessions, to be initiated at discrete time $S(p)$.

During each design session, more than one module can be applied, thus simultaneously modifying several attributes. *This is the effective concept of concurrency.*

Concurrency implies time-sharing of resources, in order that, during each design session, several modules can be simultaneously applied.

Let $M(p)$ be the set of modules applied during session p (*Concurrent Modules*).

During each design session, the amount of available resources must be shared among the applied modules such that, owing to the resources shared, each module can move the attribute value towards its target.

The attribute target value will be reached at the end of a sequence of consecutive design sessions.

Note:
- $IR(k)$, the innovation rate of the attribute state, per unit time and unit resource, which can be applied by module k;
- $XT(k)$, the desired target value of the attribute state k.

During each session p, the _Concurrency Condition_ represents the set of design modules applied during this session:

$$M(p+1) = M(p) + \{k.s.t.\, S(p+1) = A(k)\} -$$
$$\{k.s.t.\, X(k, p+1) = XT(k).\, AND.\, X(k, p) < XT(k)\} \qquad (3)$$
$$M(0) = \{0\}$$

This condition shows that each session starts when one of the two following *events* occurs:

(E1): activation of a new design module at times _A(k), which are original unknown variables to be scheduled;_

(E2): disactivation of a module if the target value of an attribute state has been reached.

During each session p, the _Condition of Simultaneous Utilization of Resources_ represents the sharing of resources among modules to be used during said session:

$$\text{if} \qquad k \in M(p) \qquad \text{then} \qquad R(k, p) \neq 0 \qquad (4)$$
$$\text{otherwise} \qquad R(k, p) = 0$$

$$\text{and in every session} \qquad \sum_{k \in M(p)} R(k, p) \leq R \qquad (4\text{-}b)$$

R(k,p), for any k,p, are the other original unknown variables, to be planned.

The *Conditions for a New Session Activation* allow to compute the time at which a new session *(p+1)* will be initiated, through recognition of the next possible event:

- estimated times at which the target state should be reached maintaining the same resource sharing:

$$\forall k \in M(p) \Rightarrow F(k,p) = \frac{XT(k) - X(k,p)}{IR(k)R(k,p)} \tag{5}$$

- the next time at which a state will reach its target value, maintaining the same resource sharing:

$$F(p) = \min_{k \in M(p)} \{F(k,p)\} \tag{6}$$

- the next time at which the activation of a new module is expected:

$$AP(p) = \min_{k \notin M(p)} \{A(k).s.t.A(k) > S(p)\} \tag{7}$$

- activation time of the new session:

$$S(p+1) = S(p) + \min\{AP(p); F(p) + S(p)\} \tag{8}$$
$$S(1) = 0$$

The consequent *Condition of State Innovation* describes how an attribute state, starting from an initial value, evolves owing to the design activity performed by the related module:

$$X(k,p+1) = X(k,p) + IR(k)R(k,p)L(p) \tag{9}$$
$$X(k,0) = X^0(k)$$

where

$$L(p) = S(p+1) - S(p) \tag{10}$$

Assuming to evaluate the different design activities to be performed by design modules, and then to represent the corresponding Gantt diagram, it follows that:
- the Gantt diagram can be obtained by <u>simulating</u> conditions (3) to (10), starting from initial values *M(0)* and *S(1)*;

- simulation of conditions (3) to (10) requires knowing the original variables $A(k)$ and $R(k,p)$;
- the original unknown variables $A(.)$ and $R(.)$ must be searched in such a way to assure the best possible design program and to satisfy the effectiveness conditions (1) - (2).

These conditions show the necessity of stating a *design optimization problem*, by selecting a suitable efficiency index.

Typical efficiency index in a design process is the design make-span:

$$J = \underset{k}{Max}\{Z(k)\} \tag{11}$$

defined by

$$Z(k) = A(k) + \sum_{p \in Q(k)} L(p) \tag{12}$$

$$Q(k) = \{p.s.t.k \in M(p)\}. \tag{13}$$

Then, planning the ILS Design Process means to solve the problem of searching for the values of $A(.)$ and $R(.)$ in such a way that the measure (11) of the design process efficiency could be optimized.

The formal statement of the ILS Design Process Planning problem is:

Find the minimum value of the design make span J of (11)
with respect to:
- the times at which utilization of each design module k must be initiated, i.e. $A(k)$;
- the rates according to which each design resource k must be utilized during each design session p, i.e., $R(k,p)$,
- in such a way that:
    - the constraints on event propagation (3) to (10) , and
    - the constraint (1) on the design program start-up time, and
    - the constraints specifying the design completion (2), (12), (13) be satisfied.

The proposed ILS Design Process Planning model appears to be an interesting tool for approaching the design problem considered and for formulating the problem constraints and goals.

Owing to the simplicity of its formulation and the opportunity of deriving constraints directly from the logical description of the system to be designed and of the resources and tools to be utilized, the introduced model allows to derive some general considerations.

First, *the proposed ILS Design Process Planning model is the effective model of a general Concurrent Engineering (CE) procedure*, because it clearly shows the time-shared utilization of resources. Conditions (4) to (10) emphasize that concurrency implies optimization of the resource utilization, both in time and in performing the different design tasks (i.e. each attribute innovation). Then, implementation of an effective CE procedure in organizing a design process implies solving a "discrete-event dynamic optimization problem", possibly finding approximated suboptimal strategies of resources utilization.

Second, *an effective CE procedure is the most general method of solving a design process planning problem*. This statement is justified by noting that CE means concurrent utilization of resources through time-sharing. This characteristic no longer appears in the case constraints of type (4) on available resources are not considered. When no limitation on resource utilization can be assumed, the design problem reduces the problem of selecting the best schedule of activities (i.e., individual module utilization intervals) over the set of admissible orderings. By representing this set in terms of graph, this is the so called PERT approach.

5. FINAL REMARKS

It is always difficult to identify and also focus on topics for recommended research activities in the near future. But sometimes real world can propose problems for research and development with a pressure for receiving practically utilizable solutions that the R&D environment must take special attention to them. In the logistic area, this is just the case.

Several problems of logistic systems design in the industrial world are currently approached by heuristic experience-based reasonings. Solutions proposed for such problems appear to be very poor. We have focused on two of these problems, on which industrial interest is particularly relevant.

Availability of methods to design a modular ILS consisting of an organization of intelligent service units can offer significant help in managing manufacturing plants in rapidly changing markets of new products and new technologies. In fact, the main drawback of innovation programs (which in turn are required to be continuously active, owing to continuous progress) is idle time forced in the plant. Modularity can reduce the plant idle time by concentrating stops of production into small areas and assuring continuity of production flows along alternative routes.

On the other hand, availability of a methodology to manage the activities in a design process can offer a relevant tool for reducing inefficiency in the utilization of design resources thus allowing reduction of time and cost of innovations.

ACKNOWLEDGMENTS

This research has been developed under the support of Consiglio Nazionale delle Ricerche, grant no. 93.03554.CT11.

REFERENCES

1. **Turnquist, M.A.**, "Manufacturing logistics for the 21st century", Transp. Res. Records No.1395, National Academy Press, Washington D.C., pp. 129-134, 1993.
2. **Dilts, D.M., Boyd, N.P., Whorms, H.H.**, "The evolution of control architectures for automated manufacturing systems", J. Manuf. Systems, Vol. 10, No. 1, pp. 79-93, 1992.
3. **Villa, A., Watanabe, T.**, "Production management: beyond dicotomy between push and pull", Computer Integrated Manufacturing Systems, Vol. 6. No. 1, pp. 53-63, 1993.
4. **Villa, A.**, "Modelling industrial logistics design", invited tutorial at EURO XIII, Glasgow, 1994 (to appear).
5. **Kusiak, A.**, (Ed.), Concurrent Engineering-Automation, Tools, and Techniques, Wiley, New York, 1993.
6. **Villa, A.** "Generalized production scheduling by a network of local intelligent controllers", Proc. 10th Int. Conf. CAPE 94, pp. 487-495, 1994.
7. **Ho, Y.C., Cao, X.R.**, Perturbation Analysis of Discrete Event Dynamic Systems, Kluwer Academic Publ., 1991.
8. **Browne, J.**, Production Management Systems, Addison-Wesley, 1988.
9. **Vamos, T.**, "Cooperative systems: an evolutionary perspective", IEEE Control System Magazine, Vol. 3, pp. 9-14, 1983.
10. **Hatvany, J.**,"Intelligence and cooperation in hierarchical manufacturing systems", Proc. 16th CIRP Int. Seminar on Manufacturing Systems, Tokyo, 1984.
11. **Joannis, R., Krieger, M.**,"Object-oriented approach to the specification of manufacturing systems" Computer Integrated Manufacturing Systems, Vol. 5, pp. 133-143, 1992.
12. **Yuh-jiu, Lin, G., Solberg, J.J.**,"Integrated shop floor control using autonomous agents", IIE Transactions, Vol. 24, pp. 57-71, 1992.
13. **Newman, P.A., Kempf, K.G.**, "Opportunistic scheduling for robotic machine tending", Conf. Artif. Intel. Appl., Miami, 1985.

14. **Saridis, G.N., Valavanis, K.P.,** "Information theoretic approach to knowledge engineering and intelligent machines", Proc. 1985 American Control Conference, Boston, pp. 1098-1103, 1985.
15. **Baker, A.D.,** "Complete manufacturing control using contract net: a simulation study", Int. Conf. Computer Integrated Manufacturing, Rensselaer Pol. Inst., Troy, N.Y., 1988

Chapter 2

TIME-BASED APPROACH IN DESIGNING AND VALIDATING ORDER BOUND LOGISTICS SYSTEMS

Marko Luhtala and **Eero Eloranta**
Institute of Industrial Automation
Helsinki University of Technology, Finland

1. INTRODUCTION

The nature of managing logistics differs depending on the kind of logistics network. It is possible to make a division between *converging* and *diverging* logistics. Converging logistics is typical in an order-bound business where the products are usually one-of-a-kind, that is make-to-order, or assembly-to-order products. In converging logistics the product is manufactured and assembled in several locations and finally delivered to a specific customer. The elevator business is a good example of converging logistics. An example of diverging logistics is the consumer goods business where the products are manufactured in a few factories and then delivered to a great number of people through a distribution network. Sharman in [1] suggests a corresponding division to logistics by emphasizing the meaning of order-penetration (OP) point in supply chain management. Figure 1 outlines some features of converging and diverging networks. Converging network is often called a supply chain and diverging a delivery or a distribution chain.

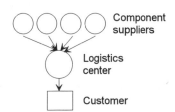

Diverging	Converging
* distribution oriented	* customer order oriented
* distribution costs 10-30 %	* distribution costs 2-5 %
* standard products	* engineering, manufacturing, delivery
* high volumes	* low volumes
* make-to-stock	* make-to-order, assembly-to-order
* consumer goods	* investments goods

Factory · Warehouses · Distributors

Component suppliers · Logistics center · Customer

Figure 1. Converging and diverging logistics network.

2. CONVERGING VERSUS DIVERGING LOGISTICS

The development of logistics research has followed the law of demand and supply. Management of diverging supply chains has developed faster than converging because the companies involved are producing high volume standard consumer goods. According to Bowersox [2] logistics in consumer goods began in the 1950s. Fast and cost effective distribution of these products is one of the most important competitive factors in that business. Moreover, some products like foodstuffs require fast and frequent delivery chains to stay current. These reasons have lead to a situation where the current logistics research is, for the most part, not an extension to production planning and control of factories, but an application of operational research (OR). The competence in converging supply chains has traditionally been based on technical superiority. Low volumes and customer specific products have kept the focus of development in the individual members of the chain.

Today, when the converging supply chains are expanding to the global scale and the need of being fast and cost effective is also increasing, the order-bound companies are eager to learn the management of supply chains. Because logistics has become synonymous to supply chain management, companies are trying to apply existing methods that are developed for diverging logistics. These methods, however, offer a marginal potential for improving converging supply chains.

In diverging logistics, issues of route optimization and warehouse location are valid due to high delivery costs relative to the price of the product. A survey by Kearney [3] indicated that distribution costs, including transportation, warehousing, inventory carrying, and administration, were an average 21% of the value added. Manufacturing is usually make-to-stock production where the lot sizes are big, product proliferation low, and process fixed. Customer related specifications are usually variations in packing or in product mix — seldom in the actual product. Manufacturing is planned and purchasing done based on demand forecasts.

In converging logistics the transportation costs are usually relatively low — in many cases less than 5% of the total product price. In this situation it is not relevant to optimize routes but to guarantee the smooth flow of order through the whole supply chain. The chain often consists of up to 6-8 vertical levels of companies. The aim is to synchronize the *diverging information flows* and the *converging material flows* and to operate as fast and as economically as possible. Demand forecasts are only used in allocating capacity — not in planning the production.

The challenge of managing an order-bound logistics system receives increasing attention from the research results, according to which customized products seem to have a highly negative correlation with ROI

(PIMS-study: see Table I) [4]. We claim that it is the poor management of converging supply chains that plays a big role in this situation — not so much the difficulties in engineering and manufacturing. If this is true, we believe that the profitability of order-bound business and customized products can be improved significantly.

Table 1. Multiple Regression Equations for ROI [4].

Profit Influences	North America	Other Countries
Real market growth rate	0.11	0.36
Unionization %	-0.07	-0.10
Exports-imports %	0.15	-0.27
Customized products	-4.01	-5.55
Market share	0.30	0.26
Relative quality	0.10	0.19
% New products/services	-0.09	-0.17
Marketing, % of sales	-0.61	-0.42
Inventory, % sales	-0.49	-0.53
Fixed capital intensity	-0,54	-0.61
Plans newness	0.04	0.17
Capacity utilization %	0.31	0.28
Vertical integration	0.29	0.24
Number of cases	1902	412

3. FROM A PHYSICAL FACTORY TOWARDS A LOGICAL FACTORY

Managing a converging supply network does not differ significantly from managing an order-bound factory. A typical structure of a global supply network consists of local front-end service units (sales, installation, maintenance), logistics centers divided on a regional, product, or customer group basis, and global components factories (Figure 2). Some people call it *globalization* because this structure aims at good flexibility and local knowledge of the customer and at the same time utilizes the economics of scale in global manufacturing.

This structure calls to mind the layout of a product-oriented factory (Figure 3). Component factories are manufacturing cells, producing components for the products. One component cell usually manufactures only one component to gain the economics of scale. Logistics centers are engineering and assembly cells, operating on a pull principle, ordering required parts from the manufacturing cells. Assembly cells are divided on the area or customer group basis. Local sales and installation units are sales representatives, sending orders to the corresponding assembly cells. Production flow analysis (PFA) by Burbidge [5] utilizes the same analogy. PFA applies similar methods in simplifying material flows between the factories (Company Flow Analysis) and inside a factory (Factory Flow Analysis).

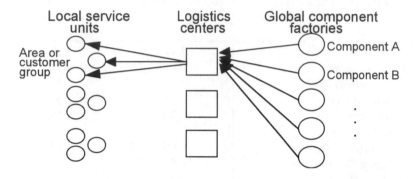

Figure 2. A typical structure of a global converging supply network.

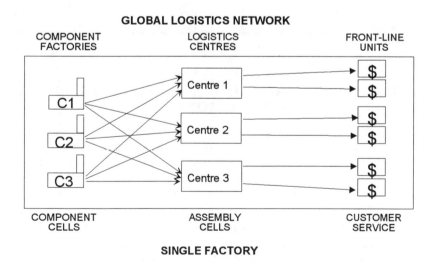

Figure 3. Analogy between a logistics network and a factory.

The analogy between a supply network and a factory can be widened even more. We suggest that a factory and a supply network have the same operative objectives. These objectives addressed by several authors are (for instance [6], [7]):

- short delivery time   - high punctuality   - high quality
- low inventories   - high flexibility   - low operative costs.

Thus, the experiences and the methods for developing control practices in factories are, for the most part, directly applicable to the converging supply networks. Concepts like JIT, streamlining, process-versus product-oriented layout, and pull principle are valid also in designing a global supply network in a customer-oriented business.

Why then is the management of a converging supply chain so difficult and inefficient? The difficulties arise from the development of a logistics network. The development process starts from a single factory that operates in local markets and the outcome is a global logistics network that operates like a well organized, but widely spread factory. Figure 4 outlines the typical path of development or "evolution" of a logistics network from a single factory to a global supply network. The time scale illustrates a possible duration of this development. Whether this path is the best one or the most efficient one is not obvious, but based on several case studies it seems to be a natural, and thus also a feasible way of development.

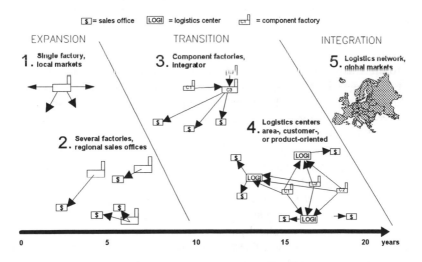

Figure 4. A typical path of development of a logistics network.

Referring to the analogy between a factory and a supply network the development process is actually a development loop where a physical factory becomes a logical factory by restructuring its operations to a global scale. In this loop the supply network goes through a heavy transformation process including both the topology (the physical structure) and the control practices of the chain. Because many of the converging logistics systems are under this transformation process at the moment, it may appear that the low profitability is a problem caused by the customer focus more than a managerial problem relating to the globalization process.

The crucial period in the development is the *transition phase* which includes arranging the independent factories into component factories, logistics centers, and front-end units. The process corresponds to changing the layout of a factory from process oriented to product oriented and establishing manufacturing and assemble cells. Based on our experiences and cases reported in the literature [8] [9], the transition phase in a factory normally takes 1-3 years. A similar change on a larger scale in a logistics network can take 5-15 years.

During the transition phase the supply chain faces several problems resulting from the mismatch between structure and control. The chain in total is inefficient due to an excess of routines and buffers in the interfaces of the units, although the units themselves would seem to be in good shape. A good example is the inventory buffers between the units. Turnover rate in a supplier's product stock and in a customer's material stock may be satisfactory when measured separately even though they both contain the same material. It is like building a wall in the middle of the warehouse to cut the inventory in half. If you can see the other side only, you may be happy with the result. A reason for extra buffers is suggested by Towill [10]. Each level in the supply chain carries additional inventory because companies in the chain get together to develop an integrated systems approach. Consequently, everyone holds buffer inventories against the very same contingencies that take place in a dynamic supply chain.

Before the transition the arguments for development and investments also arise from the local needs due to the lack of systematic understanding of the whole. Typical symptoms for this stage are long delivery times, unpredictable delivery punctuality, high occurrence of order changes, and highly variable performance in all operations. Although the savings potential in the distribution costs is only 1-2% of the net sales, the total potential is easily 10-15% as will be illustrated later with a case example. The result is bad profitability as was indicated in the PIMS-study.

These problems are a natural part of the transition process. They can be regarded as inherent problems that have to be overcome. For instance, order changes are exceptionally frequent during the transition phase. This does not indicate that the customers are changing their behavior but that many of the occurring changes are due to internal causes. In a component factory belonging to a global logistics network about 60-70% of the order changes were of internal origin. Order changes, however, are likely to decrease radically after transition phase as the internal causes have been eliminated and the network as a whole operates faster and more efficiently.

4. TIME-BASED TRANSITION MANAGEMENT

Based on a number of case studies, a focus on delivery time aspect has proven to be the best practice for managing the transition phase. Time has several advantages in matching structure and control aspects together. First of all, delivery time is one of the top three competitive factors in a customer oriented business (price, quality, and speed) (e.g., [11]). Moreover, it is probably the factor that has the biggest variance between different competitors and thus the biggest potential for gaining a competitive edge. Short lead time also indicates good internal efficiency by means of low inventories and lean processes. However, the greatest advantage of focusing on time is the ability to describe the logistics flows [12].

The supply chain is like a pipeline where orders flow downstream and become the desired products and services. If the pipeline is in good shape, the orders are in a continuous flow. Stops, bottlenecks, and unnecessary detours immediately affect the expedition of the order flow. In a supply chain there are stocks, buffers, slow processes, and unnecessary operations holding up the progress of the order. These mismatches between structure and control can be detected through delivery time analysis. Thus, delivery time and its structure are indicators of the controllability and the level of streamlining in the whole supply chain [12].

The first benefit of a time-based approach in transition management is in the definition of the supply chain. Although it may sound trivial, the key milestones of the delivery process are often hard to define. Problems arise when separating internal and external deadlines, analyzing tasks of a different nature together (engineering, manufacturing, installation), and defining the key components of the product and main operations of the delivery process for the analysis. Measuring delivery times also forces the company to divide its products into homogenous groups to get reasonable results. Some companies have established a milestone map as a part of the project planning and control procedures of the delivery. When closing the deal with the customer the dates for the key milestones are also defined (e.g., engineering/assembly/installation completed, final delivery).

The benefit of using time as a measure derives from its global and impartial nature. An hour is the same in Finland and Italy; it is also the same for engineering and welding; and the same for a managing director and a blue collar worker. When comparing historic data with todays delivery times one can ignore devaluations and floating currencies: the value of time remains the same, unlike the value of the currencies. Time can also describe both successive and parallel processes and it forms a continuum throughout the supply chain. These properties are vitally needed during the transition phase, because the problems that are being solved are mostly human related and dependent on the personal comprehensions, attitudes, and opinions of the actors in the delivery process. Time is an unambiguous tool that can be used in finding a consensus between different groups of interest and in stating arguments for the procedures needed to manage the transition phase. A time-based approach has more than once produced facts that have broken gridlocks hindering further steps of development.

In addition to locating and prioritizing the problems in the supply chain, time-based analysis has been used successfully in estimating the cost saving potential, and in allocating and monitoring investments. During the transition phase all the development resources should be directed towards a fast and painless integration of structure and control. Without the delivery time analysis uncovering the mismatches and bottlenecks in the delivery process, there is a big risk of suboptimizing the development efforts. The outcome is not only a lower return of investment, but also a longer transition phase causing operative problems to the business.

Time also suits well for benchmarking, especially for a company that has several parallel supply chains. A few companies have applied time-based benchmarking to find the best practices for delivering customized products through a logistics network. Stalk [13] points out that the best practices are easy to locate and also easy to distinguish because time relates performance directly to the actual activities.

The evolution of a logistics network is exothermic. In other words, the transition phase generates more cost savings than investments when well managed. Organizing the operations in a new way and removing "the fat" from the delivery process can offer saving potentials of up to 10-20% of net sales. In a supply chain of $24 million net sales going through a transition phase, there was the following potential identified [12]:

- Capital released from inventories: over $3 million dollars. Yearly savings of lower WIP: more than $400 thousand. Excessive level of inventory was the result of the lack of trust between the logistics center and the installation units. Huge material buffers were reserved in the installation premises to compensate for the poor punctuality of manufacturing. However, the fear of inaccuracy was exaggerated and moreover partly caused by the false ordering practices of the installation unit.
- Time related quality costs: 10% of the product price. Savings $1.6 million. Most of the quality costs occur during shipping and storing. Due to inconsistent ordering practices and a great number of rushed orders, deliveries were often incomplete (one out of three deliveries).

Also, the storage facilities in the customers premises were poor. Due to long waiting times, the product inventories often suffered damages and required personnel to take care of the quality problems.

- Productivity gap in final installation: savings $1.8 million. The lack of project planning methods in allocating workers to different projects caused excess work in the final installation of the product. Required installation hours per project were about 50% more than in other regions. The reference value for the installation hours was found through *internal benchmarking* between parallel supply chains.

The annual savings potential in this particular supply chain was $3.8 million, about 15% of the net sales. The savings were not compared to the optimal performance but to a realistic level possible to attain in a few years. In figures, it meant cutting the delivery time from 33 weeks to 18 weeks. The means for improvement were identified as training, creation of mutual trust, and development of tools for monitoring the delivery process (time-based monitoring system). As one can see, the potential was in the units close to the end customer (final installation). This is a very typical situation in a customer oriented business: the supply chain operates on pull control and thus the biggest effect on the supply chains performance is in the actions of the downstream units. Somehow this trivial sounding finding is usually not acknowledged until a time-based analysis of the chain in the transition phase is performed. Too often the development efforts are targeted by default at the upstream units (manufacturing, logistics center). The practices of the customer service unit are taken for granted even though they may cause severe problems to the rest of the supply chain. These self-induced troubles are too often solved through additional investments before finding out the true reasons for them.

5. USING *TIME PROFIT* CONCEPT IN DESIGNING A LOGISTICS NETWORK

The equation of profit based on market price and production costs is a well known concept that has been used in illustrating different pricing principles and changes in the business environment. A similar concept is available for the management of logistics in the transition phase.

The analogy of *time profit* and monetary profit is presented in Figure 5. As in pricing, the fixed factor in delivery time management is usually the market time. This is true more and more often in a customer oriented business where the customers have less tolerance for long delivery times. Companies have to define standard delivery times for different products according to the level of customizing.

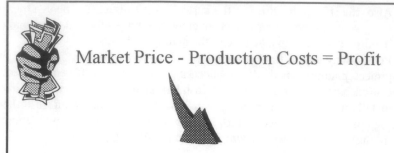

Figure 5. Concept of market time and time profit.

The time that is leftover from the delivery process, i.e., market time minus production time, is called time profit. If the production lead time is less than the required time by the customer, time profit is positive and can be used either in improving customer satisfaction or internal efficiency. In other words, the slack time can be used in generating more sales or saving costs.

Customer satisfaction and competitive edge can be improved by *cutting delivery times, offering more and faster customizing, or maximizing delivery punctuality*. Internal efficiency can be improved by using the extra time to *balance the material and information flows* inside the unit. If the time profit is negative, the company has to build inventories and invest in the production process to compensate for the long production times. This causes extra costs and thus negative time profit corresponds to monetary loss at the operative level. This is a typical situation in diverging logistics where the products are more or less standard and offered off-the-shelf through a dealer network. Customers require the product immediately which defines the market time as zero. Material stocks are kept in order to be able to deliver the products. The longer the production lead time is, the larger stocks the company has to maintain.

All the slack time in the delivery process can not be called time profit. Depending on the location and length, time buffers can also cause costs without bringing any advantage. Slack time is profitable if it is used to increase customer satisfaction or internal efficiency in the way described earlier. Any other slack time is "fat" in the supply chain. Using the time profit concept, the management of the transition phase in converging logistics becomes a puzzle to locate and measure time buffers. In the puzzle these buffers are either left as they are, transferred to another place in the supply chain, or cut to speed up the process. The outcome of this puzzle is the integration of the topology and the control methods, i.e., the objective of the transition phase is achieved.

During the design process, the operations performed in this puzzle are assessed by their effects on the operative goals like delivery time, inventory turnover, and punctuality. The effects can also be measured in money, but it is not always easy nor vital. The main issue is the awareness of the positive development that has taken place in the transition phase. Schonberger [14] outlines the same by proposing the companies to attack on waste of costing the cost reduction. Deming [15] adds that many figures that one needs for management are anyhow unknown or unknowable. As an example he mentions the effect of happier customer on sales.

An example of the time profit puzzle is illustrated in Figure 6.

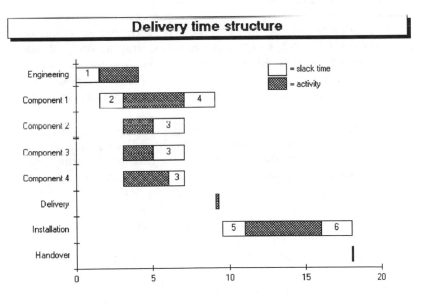

Figure 6. An example of the time profit puzzle.

Explanations to time buffers in figure 6:

1. *Slack time before engineering.* OK, if needed in leveling the capacity. Do not tie up capital.
2. *Components already ordered during engineering.* A good example of simultaneous engineering.
3. *Component WIP in the logistics center.* Components should have been ordered to arrive simultaneously. False ordering practices or lack of trust in component factories? Placing the slack time before ordering can partly reduce the need for slack time # 1. Component factory # 1 is the bottleneck: develop.
4. *Time buffer before delivery.* Unnecessary but typical — and also expensive — slack time. Schedules often include time buffers even if there was no need for them. Indicates a lack of trust in the punctuality of the component factories. Less order changes would occur if the time from order to delivery was shorter. Get rid of it!
5. *Same as # 4.* In addition to controllability problems also bad customer service if the installation takes place in customer's premises. Typical double buffer between two units (slack times # 4 and # 5). Get rid of it!
6. *Slack time before handover.* Pull control planned incorrectly. Tied capital at a maximum level. Get rid of it!

The time profit potential in the example is seven time units out of 18 (slack times # 1, # 4, # 5, # 6). The delivery time structure does not indicate whether the 18 weeks is the market time or not. The use of time profit potential on customer satisfaction or internal efficiency depends on that information. In any case, corresponding production process improvement would require astronomical investments.

Like many other development methods, the time profit approach follows the principle of incremental improvement (Kaizen) suggested by several authors (see e.g., [16]). The purpose is to use iteration in sorting out the time buffers and in integrating structure and control methods. This can be done by cutting the time span of network control gradually, e.g., by moving from the original weekly based to daily based control. The same scale is applied to all operations including promised delivery times, production planning, and delivery monitoring.

The idea can be illustrated with the following example. Cutting delivery time from four weeks to one is possible, but cutting it from one week to zero is impossible. On the other hand, cutting delivery time from seven days to four days is feasible again. In the same way, the slack times and the time profit puzzle are focused through consecutive steps.

6. TIME-BASED MANAGEMENT AFTER TRANSITION PHASE

The payoff of the time-based approach decreases after the transition phase. Lead time remains as one of the competitive factors, but time as a focus of development efforts loses its greatest potential. The issues of customer and supplier cooperation, quality and cost management, and development of network wide information systems are stressed (see figure 7). The methods to tackle these issues are predominantly not time-based. However, the potential of the transition phase is huge, as can be seen in Appendices I-II, where two examples of before and after transition phase situations are illustrated. Appendix I presents two stages of the same supply chain and Appendix II has two parallel supply chains of the same company at different stages of the transition.

7. CONCLUSIONS

The current discipline of logistics has a gray area in the management of converging supply chains. Methods that are developed for improving distribution and warehousing offer a marginal potential in order-bound business. However, the methods that were developed and tested when restructuring factories during the last decades are applicable.

There are significant similarities between a converging supply chain and a streamlined factory. The analogy covers structural aspects, operative objectives, control methods, and also the means for managing the restructuring process. The crucial period in the development of a supply chain is the transition phase when the units of the chain are organized in a new way: to form a logical factory. This phase calls to mind the change from process layout to product layout in a single factory.

The means for managing this phase are well known from the context of cell manufacturing and JIT. Based on a number of case studies, the time-based approach has proven to be the best practice for managing transition in supply chains. Time has several advantages as a measure and a tool for matching the structure and the control practices in a supply chain. When regarding slack times in the delivery process as time profit, one can use a puzzle-like method to locate the problem areas and to design new control methods.

The benefits of time-based approach in development actions decreases after the transition phase. When the slack times are removed from the process, improvements in efficiency are gained through other methods. Nevertheless, this state is still far in the future as PIMS-study indicates. From the controllability point of view the supply chains are 10-15 years behind the current status of the factories.

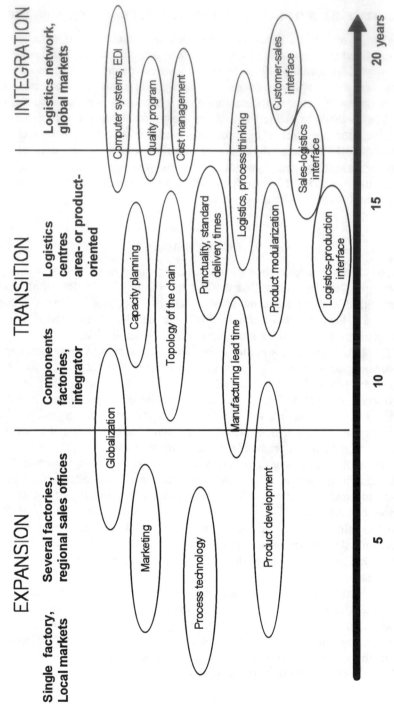

Figure 7. Management themes on the path of development of a supply chain.

REFERENCES

1. **Christopher, M. (Ed.)**: Logistics - The Strategic Issues. Chapman & Hall, 1992.
2. **Bowersox D. J.**: Emerging from the Recession: The Role of Logistical Management. *Journal of Business Logistics*. Vol. 4, No. 1, pp.21-33, 1983.
3. **Kearney, A. T.**: Logistics Productivity: the Competitive Edge in Europe. Report, 1987.
4. **Buzell, R., Bradley, G.**: The PIMS (Profit Impact of Market Strategy) Principles, Linking Strategy to Performance. Free Press, 1987.
5. **Burbidge, J. L.**: Production Flow Analysis - For Planning Group Technology. Oxford Science Publications, 1989.
6. **Hall, R. W.**: Attaining Manufacturing Excellence. Dow Jones-Irwin, 1987.
7. **Tanskanen, K.**: Supplier Management in Just-In-Time Manufacturing. Helsinki Univ. Tech., 1993.
8. **Schonberger, R. J.**: World Class Manufacturing Casebook - Implementing JIT and TQC. Free Press, 1987.
9. **Eloranta, E., Räisänen, J.**: Controllability Engineering - A Study on Developing Production Systems and Control in Finland. (in Finnish) SITRA, 1986.
10. **Towill, D. R.**: Supply chain dynamics - the change engineering challenge of the mid 1990s. *Proc Instn Mech Engrs*, Vol. 206, pp. 233-245.
11. **Gunn, T. G.**: 21st century manufacturing - Creating Winning Business Performance. Harper Business, 1992
12. **Luhtala, M.**: Performance Analysis of an International Delivery Chain, Helsinki Univ. Tech., 1992.
13. **Stalk, G. Jr., Hout, T. M.**: Competing against time - How time-based competition is reshaping global markets. Free Press, 1990.
14. **Schonberger, R. J.**: Building a chain of customers. Free Press, 1990.
15. **Deming, W. E.**: Out of the Crisis. Cambridge University Press, 1992.
16. **Imai, M.**: Kaizen. Japan Management Association, 1988.

Appendix I: An example of the transition

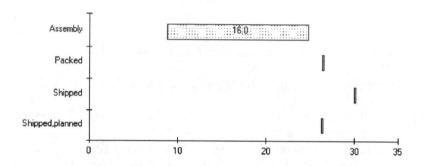

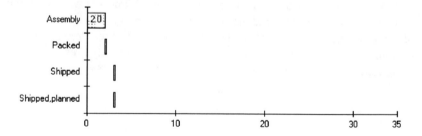

Case A. One supply chain, before and after transition phase. Standard components (assembly-to-order products).

Appendix II: An example of the transition

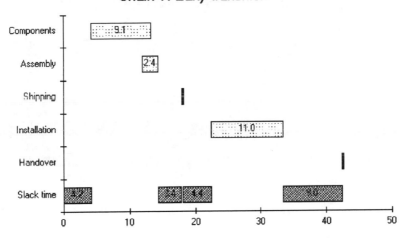

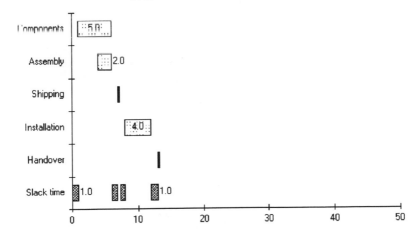

Case B. Two supply chains of the same company. Same products. Different development stages.

Chapter 3

THE FACTORY IN TRIPLICATE

Mario Lucertini
Centro Volterra and Dept. Electronic Engineering
University of Roma Tor Vergata, Italy

Daniela Telmon
Tradeoff - consulting, Roma, Italy

1. INTRODUCTION

The aim of this paper is to propose a method to represent production systems (the term production is used here in a very broad sense, although most of the quotations in the paper concern manufacturing systems) which clearly divides what the system is asked to do, from the organization of resources which is supposed to do the job, from the decision process which drives the behavior of the production system.

Most of the ideas contained in the paper are not new [1-9], but, as far as the authors know, they have never been presented in a structured and unified framework.

This attempt to represent a production system is useful if we are looking at new paradigms for strategic organizational structural changes.

Recent studies on benchmarking [10-13], on total and activity based cost methodologies [14, 15], on manufacturing systems design [16-20], on the company information flows [21, 22], stress the relevance of having a suitable representation of the system for proposing new solutions. Most of the traditional performance indicators have a restricted potential for outlining improvement strategies, the process approach has more potential, but both give a partial view of the whole system behavior [16, 18, 19, 23-25].

To have a deeper understanding of the interplay between performance indicators and system representation, the paper analyzes some examples and cases (in particular the well known push and pull production organization schemas [26]) on the ground of the representation proposed and outlines some of its possible consequences.

The need of such a structured approach depends on the extremely diverse and complicated structure of production systems. No clear line exists between small (simple) and large (complicated) systems as far as their analysis and modeling is concerned. The complicated logical structure of the decision process reflects numerous interconnections between individual facilities and plants, partly due to the flow of products and resources, partly due to the organizational links. Moreover, some of these links may change in time, as control programs are executed [27].

With so many factors involved in managing production systems to meet the marketplace demand, it is essential that a company and a plant have a clear and well-structured system which identifies documents, coordinates, and maintains all the key activities needed to assure the necessary quality actions throughout all relevant company, and plant operations. The representation of the production system used here is based on a *network of processes* to emphasize the interconnections [28].

Each process is characterized by well defined upstream processes, input, transformation process, output, and downstream processes. Inputs (e.g., materials, information) come from upstream systems or suppliers, like vendors, and outputs (products or services) are delivered to downstream systems or clients. The chart in Figure 1 summarizes this representation.

Figure 1. The transformation process.

The basic processes are aggregated and coordinated at various levels of hierarchy following the structure of the organization [29,30]. The behavior of these subsystems varies under the impact of the environment and the control instructions issued for the system to accomplish its objectives. A multilevel arrangement of the system leads to the need for sequential stage-by-stage coordination of activities within the system. The efficiency of such coordination depends on the choice of local controls at each stage, so as to ensure an optimal global performance of the system. The process of sequential coordination, conceptualized in the form of an iterative procedure, has been demonstrated to be capable in many cases of improving the global performance of a large system [31].

The basic transformation processes are generally well known and standardized, meanwhile the behavior of the whole system can be analyzed only using sophisticated modeling tools. Moreover, although the information provided by the modern information systems is, in principle, fairly complete, in practice most of the basic data and the knowledge needed to read the data is compartmentalized.

Largely, this is the nature of the technology, self-defining into narrower and narrower specialities; each advances along a cramped frontier, within strict boundaries and developing its own jargon. But it is also because the explosion of information within each field, that it becomes difficult for any specialist to explore beyond his chosen arena and to communicate across these boundaries.

Many authors conducted extensive investigations on managers who make decisions with incomplete information. Most nonprogrammed decisions involve too many variables for a thorough examination of each and managers rarely consider all possible alternatives for the solution of a problem. There exists, in every problem situation, a series of boundaries or limits that necessarily restrict the manager's picture of the world and, therefore, the decision range. Such boundaries include individual limits to any manager's knowledge of all the alternatives as well as such elements as policies, costs, and technology that cannot be changed by the decision maker. As a result, the manager seldom seeks the optimum solution but realistically attempts to reach a satisfactory solution to the problem at hand; instead of attempting to maximize, the modern manager satisfices. He examines the five or six most likely alternatives and makes a choice from among them, rather than invest the time necessary to examine thoroughly all possible alternatives. In decision making studies, many researchers have concentrated on the analysis of alternatives with given constraints. Limited attention has up to now received constraints analysis. On the other hand, constraints seem to be the real object of benchmarking [32] and the most important constraints that can be used as drivers of an improvement process depend on the embedding of a set of interconnected activities or operations in a set of interdependent organizational units.

The conceptualization proposed here is aimed to identify quantities to be measured (so that suitable performance indicators, based on such measures and on a model of the whole production process, could be evaluated), and procedures to modify the system constraints in order to obtain better performances.

It is a common experience in performance improvement [33] to see flows of resources that have been eliminated from some stage of the transformation process, pulling up at some other stage, often in a different form. Our claim is that, sometimes, this happens because the drivers we have used to realize the improvement were tailored only on one aspect of the whole system (traditionally on the organization, more recently on the transformation process) and do not integrate the information on the three main elements of the production system: the transformation process, the organization, and the decision process.

2. THE THREE LEVELS OF PRODUCTION SYSTEMS: FLOW OF INFORMATION AND MATERIALS, ORGANIZATION STRUCTURE, AND DECISION PROCESS

In the following, we propose a representation of a production system in a given time window as a set of flows (representing the process) embedded into an organization structure and driven by a sequence of decisions taken by the decision centers of the organization structure.

In this first attempt to represent the production system, we will not take into account the interaction of the organization with the environment.

2.1. PROCESS FLOWS

The process flows considered here concern what must be done by the production system and are divided into production flows and support flows.

Production flow. Flow of materials through the transformation processes. By materials we mean not only physical elements (such as, components, parts, and subassemblies), but all elements subject to the primary transformation process characterizing the production system (often in the format of information). For instance, we will consider as material the drawing developed by the engineering department, the invoices prepared by the administration department, the inventory update prepared by the materials manager. To represent the production flow, two types of information are needed on: operations (an operation is an elementary work unit of a transformation process; the set of operations required, with the corresponding precedences, gives the so called *operation graph*; operations, in a project management or activity based management terminology, are often referred to as *activities*; in production environment operations are often clustered in so called *functions*) and flows (number of units of material or information flowing through each section of the operation graph in each time interval). The nodes of the *production network* correspond to the transformation processes (and include also buffers and warehouses), the arcs correspond to the logical (often also physical) links among the nodes. The inputs of the system directly related to the production flow are sources of the production network; the outputs are sinks, and the entering flows are asked to fit the demand or, more precisely, the assigned factory workload.

Support flow. Support flows include coordination (e.g., the flow of information on the production plan and on the state of the system) and the flow of resources needed for each node of the operation graph. By resources needed we mean both the physical resources, tools, and activities needed to make the operations (e.g., machines, facilities, end effectors, time, money, maintenance, software, etc.) and the information that enable the resources distributed along the production flow (e.g., machines, material handling systems, workers) to operate, including information on who carries the output of the decision process, i.e., the flow of information that carries the decisions taken by the different decision makers (e.g., routing, scheduling, choice of operation type) to the resources in charge of performing the operations and, therefore, produces the operations (provided that the resources and the information/tools needed will be available and correct). In order to build the support flows we must quantify the work contents of each operation in the production flow. The nodes of the *support network* correspond to the transformation processes of the production network, the decision centers, the logistic facilities, the information management centers, etc.; the arcs correspond to the logical (often also physical) links among the nodes. The inputs of the system related to the support flow are sources of the support network; the outputs are typically the nodes of the production network.

It is worthwhile noticing that some flows can be considered part of the production flow or part of the support flow according to the system's representation. For example, the flow of money can be part of the support flow if we are considering a production system whose primary goal is to produce goods in given quantities, on the other hand it can be part of the support flow if we are considering a business unit whose primary goal is to produce added value.

Resources are considered both in the support flows and in the organization structure. The use of the resources for production is represented in the support flows, the supply of resources is represented in the organization structure.

2.2. ORGANIZATION STRUCTURE

The aspects of the organization structure considered focus on who is in charge of doing the activity.

Resources. Set of elements available for the production or that can be obtained from outside in the time window considered. They include all factors of production: physical resources, people, information, money, etc. Resources are the main element conditioning decisions. By definition, anything that can be used to help solve a problem is a resource, including: time, money, personnel, expertise, energy, equipment, raw materials, information.

Organizational units (OU). Set of elements of the company, each defined by its mission, goals, type of operations which is allowed to perform, type of decisions which is allowed to take, resource management strategy, etc. It could be difficult to introduce OU without considering explicitly the allocation of resources to them (in many cases the OU are considered as set of company resources), but, as our purpose is also to analyze how different resource allocation strategies influence the system performance, it is better to consider separately resources and OU.

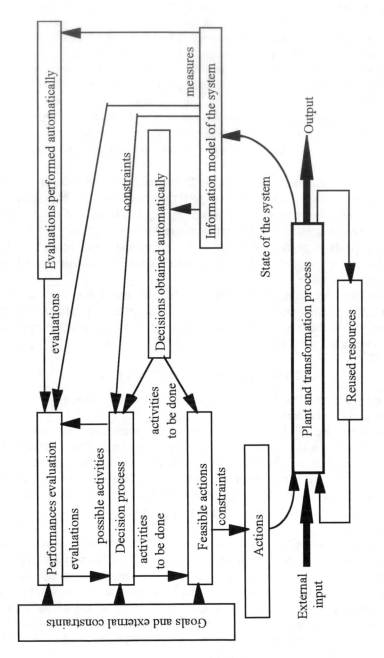

Figure 2. Automated decision.

Links. Different types of connections among the OU. They concern the way the different units interact one with the other through rules and procedures, and, in this case, they may include some decision aspects: the so called automated decisions (see Figure 2), i.e., the decisions that are taken on the ground of given rules and procedures. Depending on the system representation we are using, and on the goals of the modeling, many decisions can be included either in the rules and procedures section, or considered part of the decision process. They also concern physical links (such as transportation facilities and service networks), the information network (depending on the way the company information system has been designed and is used), the hierarchical structure (partly included in the rules and procedures section), all other forms of formal or informal communications. Some links depend only on the OU, others depend also on the resource allocation. Moreover, some links are structural, i.e., do not depend on the decisions considered here, others are the output of the decision process concerning the information exchange.

2.3. DECISION PROCESS

The flows and the organization structure interact to produce different production patterns on the ground of the decisions taken. The decisions considered are of two types: structural decisions and operations level decisions.

a) Structural decisions

Allocation of resources to organizational units. Who is allowed to operate? For each unit, the assignment gives the set of resources that are permanently assigned to it (at least in the time interval considered). There are several ways to represent the assignment. A simple and effective way is to indicate the set of operations that the resources assigned to each unit allow it to do. Let there be a bipartite graph F(O,U,E), where O is the set of operations, U the set of OU and E a set of feasibility edges between O and U (there is an edge from i to j iff unit j can perform operation i). The allocation of resources to units gives the set E.

Resource exchange patterns. For each unit, the set and the planned time schedule of the resources (in particular information) sent to and received from all other units and the environment. This is one of the core aspects of the links among OU for both physical and human resources and, above all, information. The way to deal with this part of the decision process is crucial in the system representation. In fact, although the best decisions could be made on the grounds of fairly complete information (in particular on the future), uncertainties and limited knowledge are the usual situation in the real decision environment. One of the most pernicious aspects of this process is the tendency under stress to discount the future, with the discount rate depending heavily on our emotional state at the instant of choice. Side effects are also obscure and we often neglect impacted parties who where geographically, socially, institutionally, and chronologically remote from the transaction. In a completely rational atmosphere, these conditions should prompt a search for more informations, at least to narrow the uncertainties. But there is a price in term of money and time; information is never free. The real flow of support information, the availability of resources, and the flow of decisions must be synchronized and coherent (Figure3).

b) Operations level decisions

Assignment of operations to organizational units. Who is doing what? For each unit, the assignment gives the set of operations which are assigned to it, in the time interval considered. Using the bipartite graph introduced above, the problem can be formulated as the one of finding the subset A of E, such that any operation is assigned to exactly one unit. The assignment ascribes a workload for each unit of the production system. It is called *capacitated* if any unit can perform (in the time window considered) all the operations assigned to it (without considering the delays introduced by other units or by internal scheduling problems); it is called *balanced* if the workload for each unit is proportional to their capacity, i.e., to the maximum amount of operations that the unit is able to do (with the given resources and in the time window considered). We suppose that, whenever an operation is assigned to a unit, the unit is in charge of the decisions concerning when and how to perform the operation (*how* regards the case of multiple choice in performing the operation, e.g., when two different machines belonging to the same unit can perform a same operation). The information used to make the decision depend on the information received from the other units.

Scheduling of operations in each organizational unit. When everything is done and with which resources. Aim of the decision, taken by the responsible of each unit, is to find the start and the completion time for each operation on each unit and on each resource. It is a resource constrained scheduling problem which drives the behavior of the unit.

Production and support flows represent what has to be done, the organization structure with the assigned resources and operations represents a way to realize the process but introduces relevant constraints, scheduling decisions represents a decentralized decision process in a strongly constrained environment, resource exchange decisions represent a way to make the local decisions possible and effective [34-36] (see Figure 4).

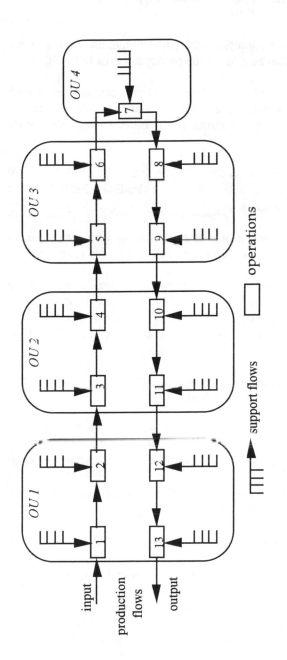

Figure 3. Organizational units in charge of production and support flows.

The organization structure produces different types of constraints for the processes:

* on the routings (a production flow can pass through an organizational unit only if the resources needed for the production exist in the unit),

* on the quantities available (an organizational unit cannot use, in each time interval and for each type of input, more resources than are available, lack of adequate resources might prove to be a significant constraint),

* on the scheduling (an operation cannot be performed by a unit before all the needed inputs are available and after a deadline after which the output is no longer usable),

* on the concurrency (different operations which must be performed, or cannot be performed, at the same time and/or by the same resource),

* on the quantities used (the quantity of a given resource used by an operation in an organizational unit, e.g., the operation time length or cost, could depend on the type of the resource used, e.g., the operator's skills, and/or on the global flow allocation, e.g., the information available),

* on the range of the decisions who are possible to take (they depend primarily on the information available).

From a modeling point of view, the boundaries include both decision variables and constraints: the decisions are taken on a subspace of the real decision space and on a subset of the complete set of the feasible decisions in the subspace.

Different authors have developed typologies of organization structures, with boundaries and constraints [3, 24, 37-40].

Among the constraints considered for improving strategies, there are, first of all, constraints on resources. These may (at a certain cost) be or may not be removed. Nonremovable constraints are usually physical, technological, or environmental constraints. The constraints on the flows of materials can be in most cases modified only with regard to wastes. Logical constraints, such as precedence and concurrency, in many cases can be removed by a different assignment of operations to OU, although with an additional cost. In fact, an important subset of logical constraints are organizational constraints, that can be often modified by suitably modifying the links among OU, e.g., company procedures.

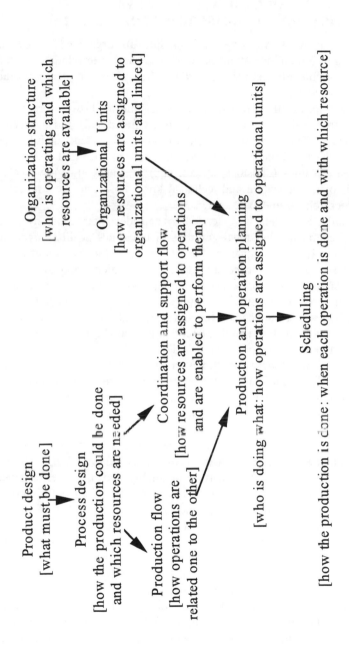

Figure 4. Flows, organization and decisions.

3. FUNCTIONAL AND PROCESS APPROACHES IN
ORGANIZATION ANALYSIS:
HOW DO YOU MEASURE PERFORMANCE?

Summing up what we have said so far, an organized system is characterized by a set of operations or activities, whose interrelations form a process, an organization structure, and the link between processes and organization structure.

There are traditionally two ways to consider and analyze an organized system: a structural (sometimes referred to as functional) and a process point of view. The structural point of view considers mainly the organization chart. All the company's resources are divided and distributed between business units (divisions, offices, etc.). Business units are created on the basis of specialization and technical know how, and are connected through a hierarchical line [33].

Traditionally, managers have tended to consider their company as a group of separated functional entities (sometimes called pyramid organization). It is the consequence of an old model of the division of work, that has had a great influence on the way in which companies were organized. The organizational chart of a company, for instance, is divided into different production units, suggesting that each unit puts together a series of activities that can be managed and measured independently. In this way barriers are created between different units and the flow of the work performed in the whole company is ignored.

The process point of view (sometimes called arrow organization) focuses on the work performed and not on the organizational chart that manages it. The process point of view identifies the main set of activities personnel is supposed to perform in order that the company produces and sells its output. These sets of activities are known as processes. To consider the organization as a set of processes and not as a hierarchy of units is a new and important theoretical requirement. *Continuous improvement*, a central element of total quality, is the measure of quality and other performance indicators in all the company processes, in order to take actions to improve them. These indicators may be customer satisfaction, number of design errors in a month, or other indicators used to characterize a process.

Therefore, in the literature on complexity in organizations, there have traditionally been two ways to read the link between activity and organization structure. The first is the classical organization model: it starts by considering the existing organization structure, and finds the best combination between activity and organization structure, with subsequent adjustments of the latter. The second approach, followed by the total quality movement, starts with the primary process, that is to say by the set of activities that have to be performed, and builds the best organization structure around them.

One of the main contributions of the quality movement is understanding that the best solution can be found only by considering the system in an integrated way, together with a continuous redefinition of the resources used for organization, in a continuous improvement effort. The continuous improvement goal focuses on the process, so as to improve the clients satisfaction and to reduce the costs associated with the improvement of clients satisfaction. Therefore, the processes, traditionally seen only as a set of activities, are now seen as a set of activities supported by a dynamic organization structure, that follows the primary process and adapts to it, according to the dynamic organization's goals.

Starting from this analytical pattern, tools are developed to analyze the company process, analyze costs more accurately (activity based costing), and realize continuous improvement.

In order to manage a company you must manage processes; to do this, the link between operations (or activities) and costs become essential for cost management. The key to understanding cost dynamics is to establish relationships between activities and their causes, and the relations between activities and costs. To consider the organization from the process point of view enables you to manage costs, managing the activities that generate them. By changing the activities you may influence the behavior of costs.

Appropriate tools are needed to identify the process output, the process internal and external clients, the work (or activities) necessary to produce the output, and finally the input. After having identified these activities, the process is then subject to value analysis (i.e., definition of cycle time, definition of the cost of each activity, determination of value added for each activity). The aim of value analysis is the identification of activities that add time and cost to a process without adding value for the client, and eliminating the activities after having identified their causes. This is the basis for identifying nonvalue adding activities, that have to be eliminated in an improvement plan.

Many studies have been lead on process definition, but much has yet to be said and studied on organization design and management seen as a support tool for the primary process. The problem is to conceive in a new way the organization structure, passing from something that already exists and self perpetuates into something that has to be built around the primary process. In the new approach, processes become the basis for organization design and management: structures and functions become changeable and they justify their existence not only for themselves and their survival, but by a means of management of the processes.

Nonetheless, as far as performance measurement is concerned, much has been done on defining indicators for the first side of the equation (the process), but very little on the second part of it (the organization). By performance we mean the result of the management of the activities in an organization in a certain lapse of time. Performance measurement is propaedeutic to planning, as it is difficult to plan if you don't know how you perform. With performance measurement we mean the feedback and the information on the management of business activities, as compared to company strategic goals and customer satisfaction.

As shown in our model, the process to be considered is not only a production flow process but also an information flow process. How do you measure the information flow? How do you measure organization constraints? Let us briefly review the process approach and its performance indicators, in order to see all that is left out.

A set of new factors have recently questioned the conventional company performance measurement system. The quality movement has contributed to a better understanding of the interrelations between clients, company processes, and business success, and to show that traditional financial based indicators have proved to be insufficient and often also misleading.

In order to evaluate and document company performance, traditional measurement systems follow the structural and not the process point of view. If we were to find performance criteria for the organization, conventional performance criteria would be effectiveness and efficiency of the functions and units. To measure this, financial reports are elaborated monthly, with reference to single organization units, represented by a profit and loss statement, to be confronted to the unit's budget. All this information is gathered for and used by top management for unit evaluation.

Process modeling and measurement has been attempted in the process approach. Process analysis tends to lose many structural aspects; in fact, it does not take into account sufficiently that the company operates in a constrained system. Process indicators can be good for evaluating final business results, i.e., the company output, but they are often not thought of for studying how structural improvement may increase the output, e.g., when we must evaluate how a constraint affects the performance and how it can be changed.

Different performance criteria that are taken into consideration when measurement system is thought of.

A first set of criteria, such as throughput, quantity of resources and different types of operations involved, financial flows, effectiveness, efficiency, productivity, defect rate, is based on measures of the production and support flows; any implication on the organization is indirect, it looks at the effects of the organization on the flows.

A second set of company wide criteria, such as organization size, number of people and their qualification, number of different operations the organization is allowed to perform, quantity of resources available, organizational structure complexity, number of managerial levels, is based on measures of the organizational structure; any implication on the flows is indirect, and only the effects of the flows on the organization are considered.

A third set of criteria, such as response or lead time, flexibility, concurrency level, system complexity, is based on mixed measures; only in looking at the process embedded into an organization is it possible to make an evaluation, often grounded on a simulation of the whole system behavior (Table 1).

Table 1. Company wide performance indicators.

Process indicators	Performance indicators
Production flow:	throughput, number of operations for each type, defect rate.
Support flow:	quantity of resource used for each type.
Production/support flows:	effectiveness, efficiency, productivity, number of precedence constraints.
Organization structure indicators	
Resources:	organization size, quantity of resource available for each type, (number of people for each specialization, number of job titles, number of facilities for each type, quantity of financial resources,...), number of different locations.
Organizational units:	number of OU for each type, number of different operations allowed, number of task areas, differentiation vs. integration level.
Links:	number of structural links among OU, links' size, time and cost, number of managerial levels, number of meshing links, number of procedures and decision rules, number of hops for each procedure, ratio (number of links, number of OU).
Decision process and mixed indicators	
	response or lead time, flexibility, concurrency level, system complexity.

We argue that, in order to measure performance, process measures are not sufficient, and that it is equally important to measure organizational variables and the joint effect of process and organization.

In the following we propose a set of short definitions of some performance criteria, in order to point out the relations with the system representation outlined above.

Effectiveness, efficiency, and productivity are classical performance criteria which evaluate in an integrated way, production and support flows. They take into consideration the input and output of the transformation process in terms of resources used and quantities produced respectively. *Effectiveness* is an "output side" issue, and is represented by the ratio between actual output and expected output, for a given input. Effectiveness is accomplishing the "right" things, in terms of timeliness, quantity, quality, and cost, for a given amount of resources used. A selected effectiveness measure should explicitly indicate whether the organization is achieving the desired results. *Efficiency* is an "input side" issue, and is represented by the ratio between the resources planned or expected to be used and the resources actually used in producing a given output. If effectiveness is doing the right thing, efficiency is doing things right. *Expected productivity* is the ratio between the expected output and the resources expected to be consumed (expected input). *Actual productivity* is the ratio between the actual output and the resources actually consumed (actual input). Productivity measures are designed to analyze output relative to the inputs. These measures may be developed for each input (labor, capital, energy, materials, data information) or in combination of inputs. Productivity for particular types of inputs and outputs, e.g., financial flows, takes different names. *Profitability* is the measure or set of measures of the relationship between revenues (output belonging, in our representation, to the production flow) and costs (input belonging to the support flow). The same criterion for nonprofit organizations is *budgetability*, i.e., the measure or set of measures of the relationship between what you said you would do and what it would cost and what you actually did and what it actually cost.

Flexibility [41-44] can be defined as the capability of the system to adapt to different requests. It has four main dimensions [45]: operation range (how many different operations can be done), throughput range (how many different production rates are technically and economically acceptable), uniformity (how well all the different operations in the range are performed), mobility (how quickly the system switches from one operation to another). To evaluate operation range, uniformity, and mobility it is necessary, in general, to consider both the process and the organization. To evaluate the throughput range we have to also take into account the decision process; in fact, the technical and economic feasibility depends heavily on the strategic and operations level decisions.

Speed can be defined as the time needed by the system to respond to different unforeseen inputs. Time competition is often a strategic orientation for companies: it allows them to be customer oriented, to improve quality, and to manage costs strategically. But, above all, time competition focuses on process speed. Examples of time competition are: time necessary to develop the design of a new product, time to develop a software program, time of arrival of an order from the client to the person responsible to process it, lead time in production lines. Process analysis is an important technique to analyze cycle time of all company processes. Linking these methodologies to activity bases cost analysis, we obtain a combination of cost, quality, and time indicators, that allows us to give time an economic value.

As a conclusion, using our model as a reference, we present in the Table 1 a set of some possible indicators that, tentatively, may be candidates for a performance measurement system. This is nevertheless the subject for a further discussion.

4. EXAMPLES AND CASES

Several examples and cases can be brought to illustrate how the conceptualization proposed allows a deeper understanding of several production environments. In the following we will quote a few of them.

The well known push and pull systems for material flow management in manufacturing lines, can be revisited on the ground of our model.

Suppose we are dealing with a production line where each unit to be produced passes through a given sequence of stages, the same for all units (flow shop) and each managed by a stage supervisor using the information coming from other stages (and, eventually, from outside) to decide the operations to be done [8,26].

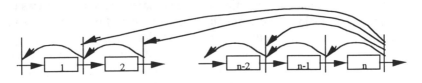

Figure 5. A push line.

A *push system* (see Figure 5) is based on the idea of computing backwards the latest start time of all operations and to perform operations as soon as the part is ready. Given a deadline for a final product, a set of deadlines for the operations needed to produce the output are derived, and, from them, a corresponding set of required ready times (or required start times) for components, features, and subassemblies, are computed for all stages. The flow of materials is managed through a global strategy based on an off-line forecast of lead times, and does not take into account explicitly local problems (e.g. delays, machine breakdown, rework, rejected parts) which can occur during the production. These local events ask for a (partial or global) complex replanning of the activities for all stages. The information needed to schedule the operations in a stage come both from downstream (deadlines) and from upstream (ready times). A push system works at each stage whenever there is a material to be worked on, with no regard to the amount of inventory downstream.

Figure 6. A pull line.

A *pull system* (see Figure 6) is based on the idea to idle a machine, even if work is waiting for it, until a signal from the buffer downstream authorizes the machine to produce. In practice, in a pull system, each stage of the production process behaves independently: as soon as the operations on an item are achieved and the output buffer of the stage is empty, the stage supervisor looks at the output of the upstream section, takes an item and starts the operations, on the ground of the information associated to it. It is a local strategy where some local problems (e.g., delays, machine breakdown) have influence only on the local processing time. The only information needed by a stage come from the upstream stage and the downstream buffer. Global performances are obtained by: globally optimizing the routing of materials and information, balancing the stages workload, managing effectively local operations to reduce setup times, eliminate rejections, standardize operations time, and resources used.

The representation of a production system proposed in section 2 allows to evidenciate the major differences between the two flow management approaches. In fact, although the production flows may be, at least in principle, very similar (e.g., same routings, same sequence of operations), the information flow network (i.e., the support flow) is completely different.

In push systems there are multiple connections among production stages to carry the information on deadlines and ready times; moreover there is a centralized information and decision management center.

In pull systems the information flow is completely embedded into the production flow and, therefore, the only connections are between pairs of adjacent stages, from upstream to downstream; moreover no online centralized information and decision management center is needed.

Several decision rules to solve the scheduling problems arising at different levels in a car manufacturing plant are compared, and the most profitable in different production scenarios, differing for production rate and mix, as well as for the number of workers and/or AGVs are pointed out [46].

For instance, the superiority of a pull-like dispatching approach has been observed over the preceding, simpler push-like system. It must be underscored that the implementation of some of these decision rules, based on optimization models, requires a change in some features of the information system (namely, the way in which parts are identified by the supervisor). Of course, the cost of these changes should be considered versus the advantages that the adoption of a new approach to material flow management would bring.

However, other decision rules were considered that do not imply major changes in the information system. In particular, a new dispatching rule, based on the conceptual model presented here and using the existing flow of information, is proposed. The rule preserves some advantages of both push and pull philosophies, without giving up simplicity (and, therefore, cost) of implementation in the existing organization.

Of course, as usual, there is a trade-off between costs and advantages which can be carefully analyzed by considering production flows, support flows, and the underlying organizational structure.

The new rule, with small and inexpensive modifications of the organization, has increased the throughput from about 1200-1300 cars per day to more then 1750 cars per day (the machine bottleneck is about 1800 cars per day), with a smaller number of AGVs.

Production management studies have recently shown that reducing lead times in manufacturing and assembly, either using automated systems, or by process analysis, is becoming insufficient [20]. In most manufacturing companies more then half of the time necessary to process an order is passed in the nonproductive sectors of a production system for reasons that are not directly connected to the transformation process. This means that a further reduction of time can possibly be achieved by changing the organization structure of the production system.

The grouping of the product in families (group technology), the sharing of resources among organizational units (with resource exchange procedures), the decomposition of the whole decision process into a sequence of smaller decision problems that can be efficiently solved, the reintegration of indirect functions in the sphere of operational personnel, the integration of all the functions of order processing in one single organizational unit (i.e., with the creation of order centers) are all solutions for improving lead times and flexibility through reorganization.

A usual approach in operations management consists of pointing out decision kernels of the manufacturing problem, with the aim of defining a sequence of subproblems (hopefully easy to solve) such that their solutions are likely to provide a good solution, although not optimal, to the global problem (such as, for instance, the above pull dispatching rule). The aim of the decomposition is mainly to recognize and isolate parts of the overall problem, corresponding to intermediate decisional problems that may be efficiently solved. An example is when we define a set of routings and a corresponding part assignment (yielding an optimal workload balance among the operating machines and minimizing the total number of part transfers) and then we solve the scheduling problem on each workstation. The solution of the routing problem yields to only partially defined production and support flows (we have not yet defined the timing of operations) but requires a fairly complete definition of the organizational structure; the second step has, in general, limited influence on the organization and vice versa.

To redesign the system we need an understanding of the relations among process and organization indicators [47,48].

An interesting application [49] concerns the sharing of tools among a set of compatible workstations or manufacturing cells. The decision involved concerns mainly the resource exchange. Tool management is a key issue in this type of cell. In fact, a machine is in a waiting state, after the execution of the first operation of a part program, whenever the tool required in order to begin one of the following operations is engaged, because it is in use on the other machine (suppose we are dealing with a two-cell system) or it is on the way in the tool handling system. In this case, a *conflict* arises in the use of a tool (or several tools used at the same time) and a decision must be made about which request must be satisfied first. Clearly, the way a conflict is solved affects the existence of the following conflicts and the system's performance. The interaction among the production flow, the flow of resources and the organization that supports such flows, can be efficiently analyzed on the ground of the approach proposed here.

A similar case [50] is the prototypal production of complex electronic systems (e.g., radar systems). Production is usually organized in cells, each in charge of producing a different subassembly. Some cells are devoted to manufacturing the mechanical subassemblies of the radar (antenna, microwave guides, rotating engines, etc.), others to the electronic components (chips). The operations inside each cell are totally ordered. Due to the integration of different technologies (mechanics and electronics), highly skilled workers (who must execute the operations) are shared among the cells. Precedence constraints exist among operations across the cell. For instance, the final phase of mechanical manufacturing can be done only after some checks of assembling feasibility between the mechanical and electronic parts are carried out. Some mechanical assembly processes are double chains, such as the parallel machining of car body and doors in automobile assembly. Doors and car bodies advance on parallel lines and, in some cases, the same robots operate on both lines to perform particular operations, such as drilling or screwing. In this case there might be no precedence constraints across the lines.

In many relevant cases the cell design problem concerns the flow management for cyclic production of one part type, where each part type must undergo a given sequence of operations, and each operation requires a given set of resources. The problem is to allocate such resources and to give a schedule of the operations, in order to maximize the throughput. This problem is considered for the assembly of a cash register, formed by several macrocomponents (a support plate, several microprocessors and memory units, a power supply unit, ...) in [51]. The assembly process of such macrocomponents has been designed for automated assembly with a robot based workstation equipped with given basic components. The synchronization of production and support flows has produced a better solution than the one proposed by plant engineers. Moreover the approach followed has allowed managers to better evaluate the different cell layouts proposed.

5. PERFORMANCE IMPROVEMENT AND TOTAL QUALITY IMPLICATIONS

In the above presentation we have not considered the total quality improvement actions and measurements. Starting from the operational definition generally used of total quality management [52], i.e., the integration of measurement and management in the five checkpoints reported in Table 2, we want to show that the indicators usually used in business practices are only a part (although the most important one) of the *company wide performance indicators* above, that could be useful according to our model.

The quality indicators most used in business practices concern basically production flows and, to a smaller extent, information flows. They evaluate the process, but do not give direct insight on how to modify the organization for quality improvement. For instance, a lower complexity and/or a greater flexibility of the organization can reduce nonconformity significantly. Lower lead times could improve customer satisfaction, the number of links between organizational units and suppliers could be a good indicator to evaluate the intervention "working with and certifying suppliers", concurrency level could be a relevant item in evaluating synergy.

In practice, while the indicators used are based on data currently available in the company, or are easy to find, most of the indicators discussed in this paper require the formulation of a quantitative model of the system and a set of numerical experiments on the ground of data that, sometimes, are not directly available with the traditional information systems.

Table 2. Quality indicators.

checkpoints	interventions and ISO 9004 corresponding sections	indicators (in business practices)
management of upstream systems	• understanding the market (7) • deploying information from functions into product design (7,8) • working with and certifying suppliers (9) • communicating requirements and expectations with suppliers (9)	• suppliers quality
incoming quality assurance	• checking if incoming products meet requirements and expectations (9) • providing employees with clear and accurate requirements	• supply non-conformity
in-progress quality management	• establishing a good understanding of the process(10) • establish process control (11) • improving process capability (10)	• process non-conformity
outgoing quality assurance	• checking that outputs meet design specifications and customer requirement (8,12) • checking if quality levels meet established goals (12)	• product non-conformity
assurance of customer satisfaction	• managing customer perceptions (7) • achieving customer satisfaction (7) • going beyond customer satisfaction • proactive assurance of customer satisfaction	• product reliability and maintenancebility • customer satisfaction
integration of quality practices	• providing synergy • coordinating quality practices within organization (4) • breaking down barriers between functional areas	

6. CONCLUSIONS

The paper focuses on the interplay between organizational structure, the decisions made by agents within the structure, the corresponding flow of materials and information, the technology and resources supporting those operations.

The conceptualization proposed looks at a factory both from the functional and the process point of view, and is aimed at identifying quantities to be measured (so that suitable performance indicators, based on such measures and on a model of the whole production process, could be evaluated), and procedures to modify the system constraints in order to obtain better performances. In fact, the process point of view, supported by the total quality movement, uses tools often insufficient to take into account and to quantify the constraints due to the organization structure.

The paper outlines the approach and a few applications, partly developed by the authors, are briefly discussed. On the basis of an analysis of these cases the approach seems promising. In fact, the conceptual framework proposed makes it possible, in several interesting cases, to obtain insights on how the system works and how it can be improved.

The approach is related to a particular perspective of benchmarking [12], to which several experiments in production research are related, e.g., TOPP program in Norway, AMICE and FTM ESPRIT programs, TOVE project in Canada, ECOGRAI project in France.

REFERENCES

1. **Bekiroglu H.** (Ed), *Computer models for production and inventory control.* Society for computer simulation, La Jolla, CA, 1984.
2. **Chang S.K.** (Ed.), *Management and office information systems,* Plenum Press, New York, 1984.
3. **Cyert R., March J.,** *A behavioral theory of the firm,* Prentice Hall, 1963.
4. **Compton W.D.** (Ed.), *Design and analysis of integrated manufacturing systems,* National Academy Press, Washington D.C., 1988.
5. **Eloranta E.** (Ed) *Advances in production management systems.* North Holland, Amsterdam, 1991.
6. **Grubbström R.W., Hinterhuber H.H.** (Eds), *Production economics.* Elsevier science, Amsterdam, 1991.
7. **Huber G.P., Glick W.H.**(Eds), *Organizational change and redesign,* Oxford Univ. Press, New York, 1993.
8. **Hirsch B., Thoben K.D.**(Eds), *"One of a kind" production: new approaches.* Elsevier science, Amsterdam, 1992.
9. **March J.** (Ed.), *Handbook of organizations.* Rand McNally, 1965.

10. **Balm G.J.**, *Benchmarking*, Qpma Press, Illinois, 1992.
11. **Lucertini M., Nicolò F., Telmon D.**, *Benchmarking and models of integration,* in: S.Y.Nof (Ed.) Information and collaboration models of integration, Kluwer Academic Pub., 1993.
12. **Lucertini M., Nicolò F., Telmon D.**, *How to improve company performances from outside: a benchmarking approach,* IFIP workshop on Benchmarking: theory and practice (invited paper), Trondheim, Norway, June 16-18, 1994.
13. **Liebfried K.H.J., McNair C.J.**, *Benchmarking: a tool for continuous improvement,* Harper Business, 1992.
14. **Johnson H.T., Kaplan R.S.**, *Relevance Lost: the rise and fall of management accounting,* Harvard Business School Press, 1987.
15. **Kaplan R.S.**(Ed.) *Measures for manufacturing excellence,* Harvard Business school series in accounting and control, 1990.
16. **Daniels A.C.**, *Performance Measurements: Improving Quality Productivity Through Positive Reinforcement,* in Performance Measurement Pub., 1989.
17. **Lynch R.L., Cross K.F.**, *Measure up: yardsticks for continuous improvement.* Basil Blackwell Inc., 1991.
18. **Maskell B.**, *Performance Measurement for World Class Manufacturing: a Model for American companies,* Productivity Press, 1992.
19. **Omachonu V.K., Davis E.M., Solo P.A.**, *Productivity measurement in contract oriented service organizations,* Int. J. of Technology management, vol 5, n.6, 703-719, 1990.
20. **Zülch G., Grobel T.**, *Simulating alternative organizational structures of production systems.* Prod. planning and control, vol.4, n.2, pp.128-138, 1993.
21. **AMICE**, *Open System Architecture for CIM,* Springer Verlag, 1989.
22. **CIMOSA**, *Open System Architecture for CIM,* ESPRIT Tec. Base Line 2.0, 1993.
23. **Arbel A., Seidmann A.**, *Performance evaluation of flexible manufacturing systems,* IEEE Tr. on System Man and Cybernetics, vol.14, 1984.
24. **Malone T., Smith S.**, *Modelling the performance of organizational structures.* Operations Research, 36, 3, pp.421-436, 1988.
25. **Suares F.F., Cusumano M.A., Fine C.H.**, *Flexibility and performance:a literature critique and strategic framework,* Working paper n. 50-91, International Center for research on the management of technology, MIT, Cambridge, 1991.
26. **Spearman M.L., Zazanis M.A.**, *Push and pull production systems: issues and comparisons,* Operations Research, 40,3, pp.521-532, 1992.

27. **Voronov A.A.**, *Management and control in large systems*. MIR Pub., Moscow, 1986.
28. **Hastings C.**, *The new organization-growing the culture of organizational networking*, McGraw Hill, 1993.
29. **Malone T.**, *Modelling coordination in organizations and markets*, Management Science, 33, 10, pp.1317-1332, 1987.
30. **Saaty v**, *Decision making for leaders: the analytical hierarchy process for decisions in a complex world*, RWS Publications, 1988.
31. **Mesaroviç M.D., Maco D., Takahara Y.**, *Theory of hierarchical multilevel systems*, MacMillan, New York, 1970.
32. **Lucertini M., Nicolò F., Telmon D.**, *Integration of benchmarking and benchmarking of integration*, J. of Production Economics (to appear).
33. **Ostrenga M.R., Ozan T.R., Harwood M.D., Mc Ilhattan R.D.**, *The Ernst and Young guide to total cost management*, Wiley, New York, 1992.
34. **Bonini C.**, *Simulation of information and decision systems in the firm*. Prentice-Hall, 1963.
35. **Huber G.P.**, *A theory of the effects of advanced information technologies on organizational design, intelligence, and decision making*. Academy of Management Review, 15, 1, pp.47-71, 1990.
36. **Kumar A., Ow P.S., Prietula M.J.**, *Organizational simulation and information systems design: an operations level example* Management Science, 39, 2, pp.218-240, 1993.
37. **Agnetis A., Lucertini M.**, *Design criteria for flexible production systems based on non-simultaneous demand models*, Proceedings of the 2nd International Conference on CIM, Troy, NY, May 1990.
38. **Baligh H., Burton R.**, *Describing and designing organization structures and processes*, Int.J.of Policy Analysis and Inf. Systems, 5, 4, pp.251-266, 1981.
39. **Carlsson B.**, *Flexibility and the theory of the firm*, International journal of production management, 3/3, 1989.
40. **Daft R.**, *Organizational theory and design*, West Pub., St.Paul, MN, 1989.
41. **Agnetis A., Lucertini M., Nicolò F.**, *Flexibility-cost tradeoff in the design of production systes: a preliminary approach*, in M.Carnevale, M.Lucertini, S.Nicosia (Eds.) Modelling the innovation, North-Holland, 1990.
42. **de Groote X.**, *The flexibility of production processes: a general framework*, Management Science, 40, 7, pp.993-945, 1994.
43. **Jae-Ho Hyun, Byong-Hun Ahn**, *A unifying framework for manufacturing flexibility*, Manufacturing Review, 5/4, dicembre 1992.
44. **Sethi A.K., Sethi S.P.**, *Flexibility in manufacturing:a survey*, The Int. J. of Flexible Manufacturing Systems, 2, 1990.

45. **Upton D.M.**, *The management of manufacturing flexibility*, California Management Review, vol. 36, n. 2, 1994.

46. **Agnetis A., Lucertini M., Nicoletti S., Nicolò F., Oriolo G., Pacifici A., Pacciarelli D., Pesaro E., Rossi F.**, *The decision process for the material flow management in a Fiat car assembly plant*, Optimization in Industry, vol.3, Wiley, New York (to appear).

47. **Lucertini M., Nicolò F., Rossetto S., Ukowich W., Villa A.**, *Models for resource allocation in the firm*. Invited paper IFAC World congress, july 19-23, Sydney, 1993.

48. **Villa A., Rossetto S., Lucertini M., Telmon D.**, *Methodological approach to planning and justifying technological innovation in manufacturing*, Computer-Integrated Manufacturing Systems, vol.4, n.4, pp.221-228, 1991.

49. **Agnetis A., Lucertini M., Nicolò F.**, *Tool handling synchronization in Flexible Manufacturing Cells*. Proceedings of the 1991 IEEE Conference on Robotics and Automation, Sacramento, CA, april 1991.

50. **Agnetis A. , Ciancimino A., Lucertini M., Pizzichella M.**, *Balancing a Flexible Line for Car Component Assembly*. Int. Journal of Production Research, 1993.

51. **Agnetis A., Lucertini M., Nicolò F.**, *Flow management in Flexible Manufacturing Cells with Pipeline operations*, Management Science, vol.39, n.3, pp. 294-306, march 1993.

52. **Sink D.S.**, *The role of measurement in achieving world class quality and productivity management*, Industrial Engineering, vol.23, n.6, 1991.

53. **Lucertini M., Telmon D.**, *Le tecnologie di gestione. I processi decisionali nelle organizzazioni integrate*. Franco Angeli, 1993.

PART 2

MODELS AND TOOLS FOR DESIGNING
INNOVATIONS IN LOGISTICS

Chapter 4

AN OPTIMAL DECISION RULE FOR REALLOCATION OF EXPERTS

Upendra Belhe and **Andrew Kusiak**
Intelligent Systems Laboratory, Department of Industrial Engineering
The University of Iowa - Iowa City, USA

1. INTRODUCTION

Design process involves both human and computer resources. Pahl and Beitz [11] suggested that, each step in the design process should be defined carefully. Similarly the types of decision that should be made and the information to be generated at each stage are also defined. An attempt is made to minimize the cost associated with a design project.

With the advent of concurrent engineering, an emphasis is given on the integration of design with manufacturing and other disciplines. Concurrent engineering is a systematic approach to product development that addresses the organizational and technological elements affecting the process of product development. It embodies team values of cooperation, trust, and sharing in such a manner that decision making proceeds with large intervals of parallel working [4].

The product development activities are performed in teams of people representing different disciplines. The project manager, who is in charge of the project acts as a decision maker (DM). The DM is responsible for scheduling and monitoring the activities to be performed by the team. The tools for management of design activities have to be more flexible than the existing project management techniques. The existing techniques are oriented towards repetitive and completely foreseen sequences of activities, which is not valid for product development (e.g., see [9]).

The DM selects members of a team based on the skills requirement of the design project. This team is identified as a core team dedicated to the development of the design. The core team performs the required design activities with the assistance provided by functional groups of the organization.

The design decisions are reached through a consensus among team members. Human skills play a major role in the design process. Skills of the individual team members along with other factors such as level of cooperation, support tools used, and so on have a direct impact on the level of efficiency of the entire team. In order to perform a set of design activities special skills may be required. Skilled and experienced individuals in the organization are usually in high demand. Due to the high cost of a unique expertise, the expert's inclusion may not be justified at all the stages of the product development.

In this paper, the design process is modeled as a discrete event dynamic system. The statement of the problem is given in Section 2. In Section 3, the measure of the design effort is defined. The discrete event model is presented in Section 4. In Section 5, the cost function for the discrete event dynamic system is developed. In Section 6, the dynamic programming algorithm to solve the decision making problem is presented. A Markov chain based approach to solve the same decision making problem is presented in Section 7. An illustrative example is presented at the end.

The objective of this paper is to demonstrate how the theory of dynamic programming and stochastic control can be used for decision making in a design process, once some of the attributes of the design process are quantified.

2. THE PROBLEM STATEMENT

For better performance of the design team, it is required that timely decisions are taken to manage the design process. Consider an expert assigned to a particular design project. Once included in the core design team the expert continues contributing to the project until the DM decides to discontinue the expert's services. The entire design process may last for N periods, where N is the maximum number of periods required to complete the design project. Since the expert's services are valuable, consider a constant cost of c units / period associated with expert's services. As the design progresses, reward is gained in terms of the amount of design effort performed in that period. At the same time another c units of cost is added to the total cost of the design project.

If the amount of design effort performed during some period k is known *a priori*, then solution to the problem of determining the time of discontinuation of the expert's services is straightforward. In practice, these quantities can, at most, be estimated with some probability. It is assumed that the probability distribution of various levels of the design effort is known.

Due to the uncertainty involved, a decision needs to be taken at the end of each period, about whether the expert should participate in the design team in the next period. The cost of including the expert in the next design period is offset by the reward gained by the design project in the form of achieving a certain design goal. The status of the design project is observed during every time period. Based on the current status of the design project, an action is taken by the DM. In this paper, the DM's decision-making process is modeled as a discrete event dynamic process. The activities of the expert are controlled over a finite horizon of the duration of design project.

3. MEASUREMENT OF DESIGN EFFORT

In order to assess the progress of the design project, it is required to measure the design effort performed during k^{th} period. The DM's decision whether to continue the expert's contribution to the design project is based on the expert efforts and usefulness of skills towards the design goal. The amount of design effort required of the expert and skills of all other persons that could substitute the expert need to be considered. Culverhouse [5] proposed a metric for design magnitude and design complexity in electronics design. Although the modified Culverhouse's measure is used in this paper, any other measure of the design effort can be used.

The proposed measure refers to the fraction of the design effort performed by an expert. The fraction of design effort performed by an expert is determined by scaling the total number of design activities involved in the design process by the skill level of the expert along with the skill level of all other persons capable of performing a set of design activities.

$$a = \frac{b}{(d + s \cdot e)} \ d \qquad\qquad (1)$$

where:

a: the amount of design effort required to be performed by the expert,
b: the number of design activities under consideration, which require a particular expertise. Since some of the activity durations may not be known, the value of b for each period is estimated at the beginning of the period.
d : the skill level of the expert in performing b activities. For each activity, the score in the form of number of similar design activities performed by the expert is determined, and d is the average of all these scores.
e : the skill level of all other persons in performing the same set of design activities. It is determined in the same way as d.
s : the percent effectiveness of the other persons' skill level as compared to the skill level of the expert.

4. DISCRETE EVENT DYNAMIC MODEL

The design process is observed at the end of periods k = 0, 1, ..., N-1 for the entire duration of the design project.

Let x_k be the amount of design effort performed by the expert (x_k is the state of the system). At the beginning, the initial state of the system, $x_0 = 0$. The value of the total design effort is estimated and then divided into m levels (scaled onto the scale of 0 to m). The amount of design effort to be performed may vary from period to period. It is assumed that these random design efforts z_0, z_1, ..., z_{N-1} are independent.

The underlying discrete event dynamic system is of the form:

$$x_{k+1} = f_k(x_k, d_k, z_k), \qquad k = 0, 1, 2, ..., N\text{-}1 \tag{2}$$

where:
k: indexes the discrete time
d_k: the decision variable to be selected with the knowledge of the state and the reward criteria
z_k: amount of design work to be performed in period k, which is a random parameter with given probability distribution
N: the horizon or the maximum number of times the decision about the expert's inclusion in the design project is to be made.

The amount of design effort to be performed by the expert, z_k, is computed at the beginning of period k. The values of z_k are obtained by using equation (1). These values of z_k are rounded off to the integer values from 0 to m. Therefore the determination of x_k automatically classifies it in one of the possible states. The possible state space for x_k consists of integer values from 0 to m.

Hence, the decision problem discussed in Section 2 is modeled as a stopping problem. Thus at each stage the DM observes the current state in the form of the design effort performed. Based on this observation and the reward criteria, the DM decides whether to continue the expert's contribution to the design project. The decision making model is shown in Figure 1.

After each observation of the system, one of the following two actions is taken:

d1: If the expert's continuation in the design project is not economical then his/her contribution to the design project is discontinued.

d2: If some gain is expected (or no loss is expected) in the next time period then the expert continues to participate in the design project.

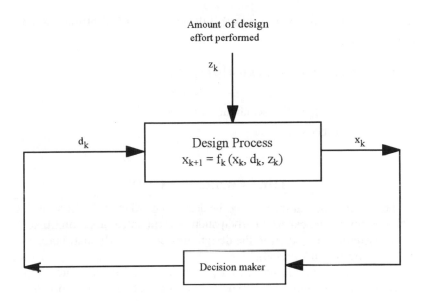

Figure 1. The decision making model.

The rule (policy) G defined below prescribes which action needs to be taken at each point of time. The concept of Markov decision process is particularly useful to model the described decision problem. It permits the use of a rule for taking an action at time k. The system equations are as follows:

$$x_{k+1} = f_k(x_k, d_k, z_k), \quad k = 0, 1, 2, ..., N-1 \tag{3}$$

$$x_0 = 0. \tag{4}$$

The finite state space is denoted by: $M = \{0, 1, ..., m\}$.

Suppose that there exist an absorbing state T (i.e., $p_{TT} = 1$), such that
$P\{x_k = T \text{ for some } k \geq 1 \mid x_0 = 0\} = 1$

Therefore:
$x_k \in M \cup \{T\}$.

The state space is augmented with an additional state (T) called as a termination state. The system moves into the termination state as soon as the decision to discontinue the expert's efforts is taken. When the system is in state $x_k \neq T$ at time k, it means that the expert is still contributing to the design project.

The function f_k is defined as follows:

$$x_{k+1} = \begin{cases} T, & \text{if} \quad d_k = d^1 \text{(discontinue)} \quad \text{or} \quad x_k = T \\ z_k, & \text{otherwise} \end{cases} \tag{5}$$

5. THE REWARD FUNCTION

The reward function for making the decision of whether to "continue" or to "discontinue" the expert's participation is formulated, assuming that the cost of assigning the expert to the design project is linearly distributed over all time periods (c per period).

The amount of design effort in a particular period is associated with a probability distribution. The expected amount of design effort to be performed in the next time period can be determined. Based on this value of the expected design effort performed in the next time period, the expected value of the reward function can be obtained.

The reward function can be written as follows:

$$E\left\{ g_N(x_N) + \sum_{k=0}^{N-1} g_k(x_k, u_k, w_k) \right\} \tag{6}$$

where,

$$g_N(x_N) = \begin{cases} x_N & \text{if} \quad x_N \neq T \\ O & \text{otherwise} \end{cases} \tag{7}$$

and

$$g_k(x_k, d_k, z_k) = \begin{cases} \dfrac{x_k}{R \cdot (N-k)} & \text{if} \quad x_k \neq T \\ O & \text{otherwise} \end{cases} \tag{8}$$

6. THE DYNAMIC PROGRAMMING ALGORITHM

The optimal stopping problem is characterized by the availability of a control, at each state, that stops the evolution of the system. Thus, in the case of design process, the DM observes the state of the system in each period and decides whether to continue the expert's contribution to the design process to gain a certain reward or to discontinue it. The reward is directly proportional to the amount of design effort performed and inversely proportional to the cost of expert's contribution for the remaining time periods, where n is the proportionality constant. The objective is to find a rule (policy) for continuing or stopping the expert's contribution that maximizes the reward. In this section, the dynamic programming algorithm to obtain such a rule is derived.

Suppose that the system is at period i. Consider the subproblem of maximizing the "reward to go" from period i to period N

$$
E\left\{ g_N(x_N) + \sum_{k=i}^{N-1} g_k(x_k, d_k, z_k) \right\}
$$

The corresponding dynamic programming algorithm over the states x_k is as follows:

$$
J_N(x_N) = \begin{cases} x_N, & \text{if } x_N \neq T \\ 0, & \text{otherwise} \end{cases}
\tag{9}
$$

$$
J_k(x_k) = \begin{cases} \max\left(\dfrac{n \cdot x_k}{c \cdot (N-k)}, E\left[J_{k+1}(x_{k+1}) \right] \right) & \text{if } x_k \neq T \\ 0, & \text{otherwise} \end{cases}
\tag{10}
$$

The function $J_k(x_k)$ denotes the expected reward at the end of period k. These functions are computed recursively backward in time.

From this algorithm the following rule (policy) G is obtained for the case when $x_k \neq T$:

An expert continues on the design project, if

$$
\frac{n \cdot x_k}{c \cdot (N-k)} > E\left\{ J_{k+1}(x_k) \right\}
\tag{11}
$$

or discontinues, if

$$\frac{n \cdot x_k}{c \cdot (N-k)} < E\{J_{k+1} \ (x_{k+1})\} \tag{12}$$

If $\dfrac{n \cdot x_k}{c \cdot (N-k)} = E\{J_{k+1} \ (x_{k+1})\}$, then both decisions "continue" and "discontinue" are optimal.

7. THE OPTIMAL STOPPING OF MARKOV CHAIN APPROACH

In this section, another formulation and solution approach to the decision making problem about the expert's contribution to the design project is presented. To solve this problem, the technique of optimal stopping of a Markov chain is used.

It is assumed that the amount of design effort performed during any period with the known probability distribution depends only on the amount of design effort performed in the previous period. This Markov chain model is used to obtain a rule that prescribes the time to stop the expert's contribution based on the reward criteria. The rule (policy) for continuing or stopping the expert's contribution is prescribed by stopping time τ.

Problem Formulation

Suppose that the process $\{z_k, k = 0, 1, ..., N\}$ is a finite state Markov chain with stationary transition probabilities $\{q_{ij}\}$, where z_k represents the level of design effort to be performed by the expert in period k.

There exist absorbing state T in the state space M, such that,

$$P\{z_k = T \ \text{for some} \ k \geq 1 \ | \ z_0 = i\} > 0 \qquad \text{for all} \ i \in M$$

The state T corresponds to the termination of the expert's contribution to the design project.

The reward measure r_k is the value associated with each state and is given as

$$r_k = \frac{k \cdot z_k}{N} \tag{13}$$

If the expert's participation is discontinued at the end of period j, then r_j is the reward gained by the expert's design effort. The aim of the decision maker is to receive the highest possible value of r_j.

The specification of stopping time, $\tau = j$ means that the expert's participation is discontinued at time j. Then the problem is to determine the stopping time τ such that, the value of $E\{r_j \mid z_0 = i\}$ is maximized.

Solution Approach

Let the function F(i) be defined as,

$$F(i) = \max E\{r_j \mid z_0 = i\} \tag{14}$$

The dynamic programming functional equations are as follows:

$$F(N) = r_N \tag{15}$$

$$F(i) = \max\left\{ r_i, \sum_j q_{ij} \, F(j) \right\}, \qquad\qquad i \in M - \{T\} \tag{16}$$

The optimal stopping time takes the form of stopping the process at those values of i where,

$$r_i \geq \sum_j q_{ij} F(j) \tag{17}$$

If F(i) were a known function, the optimal stopping time would be known.

One of the ways to obtain F(i) is to solve the system of equations (15) and (16). According to the theorem proved by Derman [7], it can be stated that,

If $\{f(i), i \in M\}$ satisfies,

$$f(N) = r_N \tag{18}$$

$$f(i) = \max\left\{ r_i, \sum_j q_{ij} \, f(j) \right\}, \qquad\qquad i \in M - \{T\} \tag{19}$$

then f(i) = F(i).

Implementation

The above solution procedures can be implemented in practice, provided that measure of the design effort is fairly accurate. The data base for all design engineers with their special skills should be maintained. All design activities can be classified under different categories based on their nature and skill requirements. When certain design activities are performed, the unbiased measure of the skills of all team members can be used to update the data base. The estimates of the probabilities can be obtained from historical data, i.e., from the similar design projects in the past.

The number of levels, m, in which the measure of the total design effort is divided, should be appropriate. For example, if m is too small then the reward measure will not change from its previous value quickly and it may result into unnecessary continuation of the expert's efforts on the design project. Similarly, if m is too large then even some temporary small changes in the value of the design effort will lead to the discontinuation of the expert's contribution.

Since the number of design activities running in a particular period and available human resources with different levels of expertise may nondeterministically vary from period to period, the above procedure is useful.

Numerical Example

Consider design of an electronic product. Suppose that the total number of design activities to be performed is 150.

There are 40 activities which need an expert with some special hardware design expertise.

The total design effort for these activities is:

$$a = \frac{40}{(4 + 0.8(5))} \cdot 4 = 20$$

where, the expert's skill level and other engineer's skill level were 4 and 5 respectively.

The effectiveness of other members' skills is only 80% of the expert's skills.

The value of amount of design effort is classified into 20 levels. Thus, the amount of design effort remaining to be performed ranges from 0 to 20 units (i.e., taking m = 20).

The design project is expected to last not more than 10 time periods.

In each period, the number of activities to be performed by the expert can be determined and the amount of design effort can be obtained.

Suppose that from the past data it is determined that the design effort required in any time period may not exceed 5 units on the scale of 0 to 20.

The amount of design effort performed in the next period can be estimated based on the probability of attaining different values (levels) of design effort in the next period.

The probability transition matrix, P, for transition from one state to the other for 5 states (i.e., from 0 to 4) is constructed. The matrix P is as follows:

$$P = \begin{bmatrix} 0.1 & 0.8 & 0.1 & 0 & 0 \\ 0.4 & 0.3 & 0.3 & 0 & 0 \\ 0 & 0 & 0.1 & 0.1 & 0.8 \\ 0 & 0.02 & 0.08 & 0.2 & 0.3 \\ 0 & 0 & 0.1 & 0.5 & 0.4 \end{bmatrix}$$

Based on this probability transition matrix, the expected value of the amount of design effort in the next period based on the actual value of the design effort in the present period, can be calculated. When it is determined that the decision of continuing the contribution of the expert in the next period may not be cost effective (based on the expected value), the process is forced into the termination state.

Suppose that the value of the proportionality constant in the reward parameter, $n=20$, and the cost per period of the expert's contribution, $c=5$.

Let $z_0 = 3$. Therefore, $x_1 = z_0 = 3$.
$J_1(x_1) = 1.33$, and $E(J_2(x_2)) = 0.99$
Since $J_1(x_1) > E(J_2(x_2))$,

the decision is to continue the expert's contribution, i.e., $d_1 = d^2$.

Let $z_1 = 4$. Therefore $x_2 = 4$.
$J_2(x_2) = 2$, and $E(J_3(x_3)) = 1.88$

$J_2(x_2) > E(J_3(x_3))$, therefore $d_2 = d^2$.
Let $z_2 = 2$. Therefore $x_3 = 2$.
$J_3(x_3) = 1.14$, and $E(J_4(x_4)) = 2.46$

$J_3(x_3) < E(J_4(x_4))$, therefore $d_3 = d^1$.

At this moment it is determined that the decision of letting the expert continuing on the design project for period 4 may not be economical. Hence, the decision is taken to discontinue the expert's contribution to the design project at the end of period 3, for the remaining periods.

8. CONCLUSION

In this paper, a metric for the amount of design magnitude was developed. This metric allowed the use of techniques like dynamic programming in the decision making process. The use of control laws throughout the decision making process made it possible to encounter uncertainty involved in the design process and reduce the cost of the expert's services. The rules for decision making are structured on the basis of transition probabilities and are independent of the metric used for the amount of design effort. The inherent properties of dynamic programming were used to express the recursive application of these rules.

ACKNOWLEDGMENT

This research has been supported by the contract No. DAAE07-93-C-R080 from The U.S. Army Tank Automotive Command.

REFERENCES

1. **Aoki, M., and Li, M.T.**, 1969, "Optimal Discrete-time Control Systems with Cost for Observation," IEEE Transactions on Automatic Control, AC-14, pp. 165-175.
2. **Bertsekas, D.P., Shreve, S.E.**, 1979, "Existence of Optimal Stationary Policies in Deterministic Optimal Control," Journal of Mathematical Analysis and Applications, Vol. 69, pp. 607 - 620.
3. **Bertsekas, D.P.**, 1987, *Dynamic Programming: Deterministic and Stochastic Models*, Prentice-Hall, Englewood Cliffs, N.J., pp. 55-102.
4. **Cleetus, K.J.**, 1992, "Definition of Concurrent Engineering," CERC Technical Report, West Virginia University.
5. **Culverhouse, P.F.**, 1993, "Four Design Routes in Electronics Engineering Product Development", Journal of Design and Manufacturing, Vol. 3, No. 2, pp. 147 - 158.
6. **Culverhouse, P.F., Ball, L., and Burton, C.J.**, 1992, "A Tool for Tracking Engineering Design in Action," Design Studies, Vol. 13, No. 1, pp. 54-71.
7. **Derman, C.**, 1970, *Finite State Markovian Decision Processes*, Academic Press, New York, NY, pp. 103-117.
8. **Kusiak, A.**, (Ed.), 1993, *Concurrent Engineering: Automation, Tools, and Techniques*, John Wiley, New York.
9. **Londono, F., Cleetus, K.J., Nichols, D.M., Iyer, S., Karandikar, H.M., Reddy, S.M., Potnis, S.M., Massey, B., Reddy, A., and Ganti, V.**, 1992, "Coordinating a Virtual Team," CERC Technical Report, West Virginia University.

10. **Nevins, J.L., and Whitney, D.E.**, 1989, *Concurrent Design of Products and Processes*, McGraw-Hill, New York.

11. **Pahl, G., and Beitz, W.**, 1988, *Engineering Design: A Systematic Approach, Design Council*, London, UK.

12. **Ross S.**, 1983, *Introduction to Stochastic Dynamic Programming*, Academic Press, New York.

13. **Ross S.**, 1970, *Applied Probability Models with Optimization Applications*, Holden-Day, San Francisco.

14. **Smallwood, R.D., and Sondik, E.J.**, 1973, "The Optimal Control of Partially Observable Markov Processes over a Finite Horizon," Operations Research, Vol. 21, No. 5, pp. 1071-1088.

15. **Thamhain, H.J.**, 1990, "Managing Technologically Innovative Team Efforts Toward New Product Success," Journal of Product Innovation Management, Vol. 7, No. 1, pp. 5-18.

Chapter 5

PETRI-NET MODELS FOR LOGISTIC PLANNING

Hans-Jörg Bullinger and **Frank Wagner**
Fraunhofer-Institut fur Arbeistwirtschaft
und Organisation - Stuttgart, Germany

1. INTRODUCTION

In recent years, the structure of markets for consumer and investment goods has changed from offering to demanding markets. For industrial manufacturing, as a consequence, specific customer wishes can only be met by flexible organization structures inside the enterprises. Shorter product-life-cycles with the referring increased planning frequencies stand for the processing of planning efforts in a great style. Flexible logistics for the process to be handled become a determining production factor and demand intensive planning. Simulation methods are well-known and often used, but using logistic net models for mathematical analysis and logistic evaluation are seldom. Petri-net models can be used for both: simulation and analysis of logistic systems.

2. DISCRETE-EVENT SYSTEMS AND PETRI-NETS

Discrete events as they occur for example during logistic and manufacturing processes — an unsteady change of state distributed over a range of time — can be depicted to a far extend with timed Petri-net models. They offer a methodology to mathematically formulate and calculate discrete events. For an introduction into the basics and the theory of Petri-nets, we refer to Brauer [1] or Reisig [2]. We recommend colored Petri-nets or predicate-transition nets instead of classical place-transition nets. Predicate-transition-nets are more compact than place-transition-nets, but can be converted to them. For example, different logistic entities (e.g., carriers) can be modeled as tokens with different predicates on a single place or as various tokens on different places of a place-transition-net. Another extension of classical Petri-nets are timed Petri-nets: classical Petri-nets follow only a causal order, independent of time. For simulation purposes, however, a timed sequence is necessary.

For logistic planning tasks more pragmatic attributes are relevant. The advantages of Petri-nets simulations against conventional simulation methods can be described as followed:

- Simplicity
  Petri-nets consist of only four elements — transitions, places, connectors, and tokens. The orientation towards a specific problem happens through the referring comments of the elements. Beginners are capable of constructing their own models after a rather short time, which is one of the prerequisites for the acceptance of a method and its tools.

- Clear Structure
  With a well thought over structuring and incorporation of graphics-oriented modeling tools, even complex systems, for example, assembly units, can be clearly described. The abstract description of a system with the help of a programming language would hinder people with little programming experience to handle the problem.

- Versatility of Analytical Methods
  For the investigation of Petri-net models there exist analytical methods, which are capable of determining special properties of the systems, for example liveliness or traps.

- Explicit Depiction of Information Flows
  In Petri-net models data, e.g., states of a machine are only available locally. Any transport of data and information must be modeled explicitly. Normally there are no global variables or data.

- Top-Down and Bottom-Up Approach
  The Top-Down approach seems to be very useful since a realistic depiction of system structures is possible with a maximum degree of integration of real world data. The reuse of already existing building blocks and modules for a Bottom-Up approach is also possible.

The use of Petri-nets as planning tools has the consequence, that these tools are not only used by specialists but also by people designing a complex logistic process. A possible modeling technique has to be clear in its structure and arrangement. Furthermore, the technique needs to be easily comprehensible. The Petri-net method leads to a modeling technique which combines these three demands. So these Petri-net based tools are ideal for modeling and planning of systems.

3. TYPICAL APPLICATION: FLEXIBLE ASSEMBLY SYSTEMS

A characteristic property of "Flexible Assembly System" is that each workpiece can pass through the system in different ways. The determination of the path through the system of every single workpiece is done by the assembly control. Therefore, different information has to be determined, transmitted, and processed. Generally, information is evaluated at distinct levels: in the level of machine control, the cell control, or the control of the complete assembly system. Hence some specific data has to be transmitted, other data are solely processed locally leading to hierarchic and decentral information processing. Most of these planning problems of Flexible Assembly Systems can be generalized to a problem of material and information logistics.

This hierarchization and decentralization of information processing and control can be simply depicted with Petri-nets. Basically all information is available locally, at their place of generation. If, for instance, the condition of a machine or the booking of buffers in a higher level has to be evaluated, the information has to be transmitted, and this transmittance has to be depicted in the model. In colored Petri-nets, tokens are marked through predicates with different values (colors), representing different information at any place of the model, e.g., consumed time and arised costs.

In order to ease the modeling of assembly systems, the use of a library containing Petri-net modules, which stand for physical units of the system, is helpful. The library contains, for example:

- sources and drains,
- assembly, manufacturing, and testing stations (manually, automatically, hybrid)
- queuing and non-queuing transport facilities
- special handling elements like lifts, multiple-vertical-transversal units, local controls for the material flow

which can be parameterized by the user. Now it is possible to model a system faster, using the previously developed or supplied modules of functional units of an assembly system. In Figure 1 the model of a flexible assembly system is depicted.

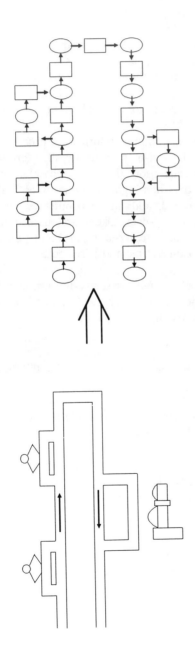

Figure 1. Petri-net modeling of an assembly system.

4. SIMULATION AND ANALYSIS

Various kinds of mathematical models and methods are used for logistic planning. Besides conventional diagram-based planning, the Operations Research methods "Simulation" and "Queuing Models" are mostly used. Petri-net models as discrete state and event models offer analytical possibilities and simulation: simulation as a numeric execution and analytical methods which are based on the mathematical foundation of Petri-nets. Both approaches have advantages and disadvantages.

Simulation as a well-known technique in logistic engineering is widely used. Many user-friendly tools and systems with graphical user-interfaces and animation features are available. But a single simulation run gives one single sample of the behavior of the system, especially when stochastic models (as in logistics) are used. Only a single trajectory in the state space is monitored. For the assessment and evaluation of the system a lot of simulation-runs (so called replications) are necessary to cover most of the possible trajectories. For a system with stochastic behavior, sophisticated statistic procedures are necessary in order to get estimations of the behaviour and performance of the investigated logistic system [3].

Analysis as the complementary approach covers the whole possible behavior of the model. All possible trajectories in the state space are covered and analysed, meaning that all possible solutions of the model are investigated. Unfortunately, the analytical procedures ignore time semantics added to the net. In the theory of Petri-nets only a causal order exists, timing does not have any meaning. So any effects arising from timing or scheduling are not taken into consideration by analytical procedures. A simple example of the application of analytical procedures to a Petri-net representing the logistics of an assembly station is presented in the next chapter. A good introduction of the application of Petri-net analysis is given by Jensen [4].

The best way of using Petri-net models for logistic planning is to apply and integrate both approaches and their results. Analysis of Petri-net models result in very intensive computations, especially if the computation is based on the reachability graph. The analytical procedures are useful for verification and validation of smaller models or for modules and building blocks, but not always for the complex and large Petri-net models [5] known in logistic planning.

5. ANALYTICAL RESULTS: AN EXAMPLE

The solid mathematical foundation of Petri-nets led to a large number of methods developed for their analysis. For example, analysis of net invariants or deadlock analysis.

In the following, selected analysis methods are applied to the Petri-net model of an assembly station (Figure 2).

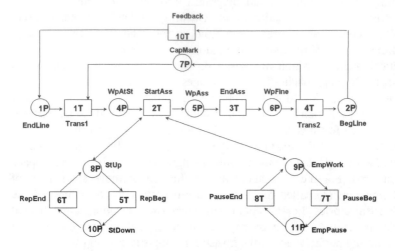

Figure 2. Petri-net model of an assembly station.

The assembly station is modeled using eleven places and ten transitions. Besides the actual assembly of a workpiece, the net models two more scenarios, namely: a repair-scenario for a station of the assembly line and an employee of the assembly line alternatively working and taking breaks. The assembly of the workpiece is covered by places 1P, 2P, 4P, 5P, 6P, and 7P, the repair-scenario by 8P and 10P and the behavior of the employee by 9P and 11P. For a full description of places and transitions refer to the following table.

EndLine: end of assembly line
WpAtSt: workpiece at station
WpAss: assembly of workpiece
WpFine: assembly of workpiece finished
BegLine: begin of assembly line
CapMark: station idle
StUp: station up, station ready to assemble
StDown: station down
EmpWork: employee is working
EmpPause: employee is taking a break

The initial marking is given by each of the places 1P, 7P, 8P, and 9P holding one token. Prior to the analysis a reachability graph of the Petri-net is needed (Figure 3). The reachability graph is a direct graph which relates the set of reachable markings (nodes in the reachability graph) to transition firing sequences (edges of the reachability graph).

For example, in Figure 3 there are 20 different reachable markings represented by 20 nodes. Node 1 reflects a marking of the Petri-net with tokens sitting in places P1, P7, P8, and P9 (indicated by the four numbers next to node 1).

Having this marking, the transitions 5T, 7T, and 1T are enabled, that is ready to fire, (indicated by outgoing arcs). If transition 1T fires, marking 2 with tokens in places P4, P8, and P9 will be the new marking.

The analysis of the Petri-net and the reachability graph led to the following results:

a) Analysis of Transition-Conflicts
    The net is not statically conflict free, meaning that there are transitions with common pre places. For example, transitions 2T and 5T are in static conflict.
    Moreover, the net is not dynamically conflict free. For example, if there is one token in 4P and one token in 8P, 2T and 5T are in dynamic conflict, meaning if 2T fires, 5T is disabled and if 5T fires, 2T is no longer enabled.

b) Analysis of Conservative Property
    The net is not conservative. That is, there are transitions consuming a number of tokens from preplaces which is different from the number of tokens produced on postplaces. For example, transition 1T consumes two tokens and produces one token.

c) Deadlock Analysis
    There are four minimal deadlocks.
    • places 8P and 10P
    • places 9P and 1P
    • places 4P, 5P, 6P, and 7P
    • places 1P, 2P, 4P, 5P, and 6P

Each of these sets of places is called deadlock, because taking away all tokens in a deadlock causes dead transitions. That is, if the tokens of a deadlock are once taken away, the deadlock will stay in this state forever. The deadlocks are called minimal, since they do not contain another deadlock.

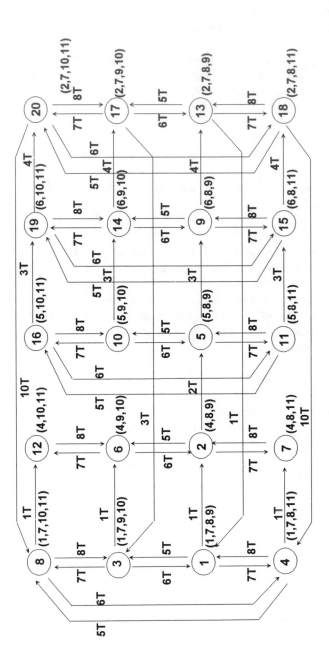

Figure 3. Reachability graph of Petri-net from Figure 2 .

d) Analysis of S-Invariants

Four S-Invariants (Place-Invariants) can be identified. They denote properties of the net independent of its dynamic behavior. The invariants are:

- $1P + 2P + 4P + 5P + 6P = const = 1$

  (the meaning of this notation is, that the sum of markings of the places 1P, 2P, 4P, 5P, and 6P stays constant/invariant). Considering this invariance, the following property can be deduced: A workpiece is either on the EndLine (4P), on the assembly line (5P), or on the BegLine (6P).

- $4P + 5P + 6P + 7P = const = 1$

  The deduced property is: There is either a workpiece assembled or there is free capacity (7P).

- $8P + 10P = const = 1$

  Deduced property: There is either a failure (10P) or not (8P).

- $9P + 11P = const = 1$

  Deduced property: The employee is either working (9P) or not (11P).

e) Analysis of T-Invariants

The net holds three T-Invariants (Transition-Invariants). T-Invariants describe firing sequence of transitions which reset the initial state of the net. In the example, to reset the initial state, the following transitions have to fire:

- 5T, 6T   or   7T, 8T   or   2T, 3T, 4T, 10T

f) Analysis of Liveness

The net is alive, meaning that the net can go into any state any time. There are no "dead ends".

g) Analysis Related to Reachability Graph

The net is reproducible. That is, the initial marking is reachable from at least one marking. For example, the initial marking, which is represented by node 1 in the reachability graph, is reachable from node 4 and node 3 by transitions 8T and 6T, respectively.

The net has no dead transition at initial marking. Each transition (1T, 5T, 7T) is enabled to fire for one reachable marking.

The net has no dead markings. That is, there is no marking reachable where no transition is enabled to fire (each node of the reachability graph has at least one outgoing arc).

The net is resetable. Differently speaking, the initial marking (node 1) is reachable from every reachable marking (nodes 2-20).

The analysis of this example was done using the commercial Petri-net analysis tool PAN [6].

6. BUSINESS PROCESS MANAGEMENT

Currently most management people are discussing the idea of Business Process Management. This can be seen as a transparent design and management of the whole business process and their costs inside the company, or in other words, as an extension of classic logistics to modern information logistics. The extended approach leads to some questions:

- How to get a clear idea of the current process with all its complexity?
- How to document, analyze and evaluate it?
- How to evaluate alternatives, different scenarios, and how to perform a necessary "what-if " analysis?
- Where and how to change the organizational and technical architecture to the new development process?
- And how to support all of this task with a comfortable software tool?

With the experience from various logistic and consulting projects and the background of modern simulation/analysis methods and systems, a methodology was developed to solve this problems [7]. The fundamental idea of the methodology is simple: "Create a dynamic model of the business processes and use it for improvements." This development of business process models must be supported by an easy, understandable method and a comfortable computer-aided tool. Figure 4 shows an example of a generic business process as part of a design process.

The same Petri-net methods and software tools are used for analysis and design of these organizational processes. In analogy to the well-known logistic process models and their simulation this idea is successfully applied to engineering, product development, and other main business processes. Petri-net based Business Process Management tools start to get more successful and popular [8]. The example shown in Figure 4 was modeled with a user-friendly Petri-net tool named PACE [9], which allows replacement of the standard Petri-net symbols by individual icons.

7. OUTLOOK: LOGISTIC PLANNING AND CHAOS THEORY

Logistic systems can be seen as mostly nonlinear systems with many degrees of freedom. Studying these logistic systems during the last years showed that their behavior depends strongly on small changes of system parameters which is typical for nonlinear systems. Research has started, whether the theory and instruments of nonlinear dynamics or popular "chaos" can be used to improve logistic planning and control [10]. First results using Petri-net models and a characterization by the Kolmogorov entropy [11] for the degree of uncertainty are quite optimistic. One of the current problems of this approach is the differentiation between stochastic and deterministic nonlinear behavior.

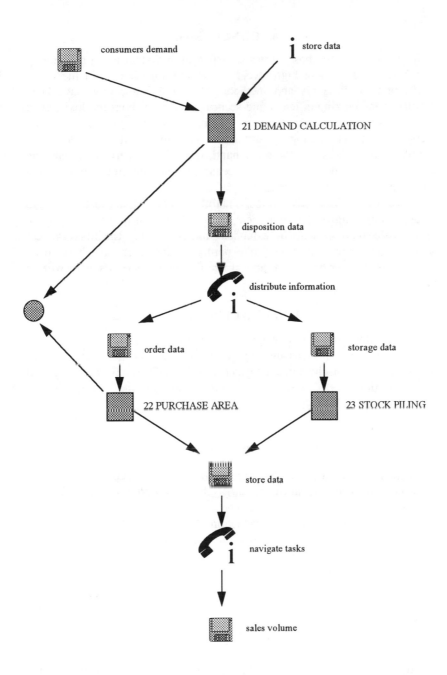

Figure 4. Petri-Net Model of a business process (with individual icons).

8. CONCLUSION

In recent years the possibilities of depicting a system using a planning tool have reached quite high levels. There are still deficits in the direct application of the tools and methods. Very few algorithms have been developed for making efficient and secure statements about the logistics of complex systems, for example, a flexible assembly process. Existing methods and tools, like logistic simulators, have to be better embedded into the planning process. On the other hand, they already serve as a valuable support tool for designing future scenarios and planning variants and alternatives in an integral manner.

Petri-net based methods and tools, mainly simulation, are just starting to be applied by industrial practitioners, enforced by their ability to use them for logistic management in extended domains, e.g., Business Process Management. New approaches, like nonlinear dynamics, still have to be investigated further before a practical use for logistic management will be available.

ACKNOWLEDGMENT

We want to thank our colleagues Mr. Wörner, Mr. Imig, and Mr. Gessl who contributed their ideas and some research work to this paper. We want to thank the Deutsche Forschungsgemeinschaft (DFG) which supported some of the work presented in this paper in the frame of the Sonderforschungsbereich 158.

REFERENCES

1. **Brauer. W., Reisig W., Rozenberg, G.** Petri Nets: Central Models and Their Properties. Advances in Petri Nets 1986, Part I. Springer. Berlin, a.o. 1987.
2. **Reisig W.,** Petri Nets. An Introduction., Springer. Berlin, a.o. 1985.
3. **Law A.M.** Simulation Modeling and Analysis. 2nd edition. Mc Graw-Hill, Inc. New York, a.o. 1991.
4. **Jensen, K.** Coloured Petri Nets. Basic Concepts, Analysis Methods and Practical Use. Volume 1. Springer. Berlin, a.o. 1992
5. **Krauth J.** Modellierung und Simulation flexibler Montagesysteme mit Petri-Netzen. OR-Spektrum. Nr. 12. 1990.
6. **PSI GmbH** NET-Analysator PAN. Release 1.0. Manual. PSI GmbH. Berlin. 1989.
7. **Bullinger H.-J., Wagner F.** A Model-Based Methodology for Management of Concurrent/Simultaneous Engineering. In: Nof, S. (Ed.): Integration: Information and Collaboration Models. Kluwer. Doordrecht. 1994.

8. **Wagner F., Fischer U.** Methoden und Werkzeuge zur Analyse und Auslegung objektorientierter Systeme. In: Bullinger, H.-J. (Ed.): Objektorientierte Informationssysteme II., Springer. Berlin, 1992

9. GELAG PACE Tool Reference Manual Version 2.1.1. Grossenbacher Electronik AG. St. Gallen. 1993.

10. **Horns A.** Job Shop Control under Influence of Chaos Phenomena. Proceedings IEEE International Symposium on Intelligent Control 1989, 227-232. IEEE Computer Society Press. Washington DC. 1989.

11. **Schuster H.G.** Deterministic Chaos - An Introduction. Physik-Verlag. Weinheim. 1984.

Chapter 6

ON SOLVING STOCHASTIC PRODUCTION PLANNING PROBLEMS VIA SCENARIO MODELING

Laureano F. Escudero
Dept. of Statistics and Operations Research
Universidad Complutense de Madrid, Madrid, Spain

Pasumarti V. Kamesan
IBM T.J. Watson Research Center
Yorktown Heights, NY, USA

1. INTRODUCTION

Many important planning problems require models that can account for the uncertainty inherent to planning. In the area of manufacturing, the planning and utilization of production capacity is one such problem. Such decisions have to be made in the face of uncertainty in several parameters, the most important being *market demand for the products being manufactured*.

Capacity planning problems are of two types. The more commonly discussed problem of deciding how much capacity to acquire and how to plan its utilization is a strategic problem that deserves careful analysis. Eppen, Martin, and Schrage [15] provide an excellent discussion on the subject. They use a scenario approach as in this paper, but their emphasis is on longer range decisions regarding facility selection for manufacturing. In the tactical time horizon, however, capacity problems are normally resolved through inventory buffers, additional workloads, or through alternate sourcing. Although new capacity cannot be acquired in this time horizon, it is often possible to develop alliances with other manufacturers or vendors to manage the production of uneven or unanticipated production volumes.

The models in this paper were developed with the tactical time horizon in mind. We consider the case where there are two sources of supply, one of which is termed *in-house production* and the other source may be referred to *as vendor supply*. Although it is assumed that there is no capacity limit on the supply from the second source, this assumption can be easily relaxed.

The problem then is to determine both in-house and vendor production plans for multiple products over a horizon of time periods with uncertainty in the demand and limited in-house production capacity. The primary contribution of this paper is in the application/practice of mathematical programming and hence the focus of the paper is on: i) identifying the relevant practical issues, ii) modeling, iii) development of a suitable computational approach, and iv) their application to the practical problem described above.

We need an approach to model the uncertainty in the problem data. The traditional approach is to make distributional assumptions, estimate the parameters from historic data, and then develop models to take the uncertainty into account. Such an approach may not be appropriate if only limited information is available. Also, in applications such as forecasting, analyzing business volumes, etc. it is often necessary and possible to take into account information that is not reflected in the historic data (e.g., new markets, new products, etc.). In the models to follow, uncertainty in the demand is captured as follows. First, a few *outlooks* of the unknown demand are created and then several scenarios (possible realizations) of demand over time are generated. For example, one may use only 3 *outlooks*, such as low demand, medium demand, and high demand and, then, generate a certain number of scenarios from these outlooks. In any case, the goal is to determine implementable policies for both in-house production and vendor sourcing that:

- take into account the possible demand scenarios.
- hedge against the uncertainty and provide for recourse actions in the future.
- and minimize the expected cost of holding inventory, in-house production, and vendor sourcing.

The idea of using scenarios to model uncertainty is a well recognized technique, at least among practitioners. The use of this technique in single period static models is straightforward. When applied to multiperiod sequential decision models, however, several interesting questions arise.

Section 3.1 discusses these questions briefly and presents a framework that uses an approach (due to Rockafellar and Wets [31]) to deal with these issues.

Our solution methodology is as follows. First, a multistage Stochastic Programming (SP) model is developed, and the concept of nonanticipative policies is described. Scenarios are used to model the multistage decision process and the recourse actions available to the decision maker are introduced.

Finally, the multistage model is approximated by a three-stage model, wherein the time periods are clustered into three decision stages. We present two alternative approaches for the Linear Programming (LP) equivalent model to the SP model, namely a *compact representation* and a *splitting variable* representation. We exploit the structure of the compact representation and develop a crash procedure to generate an advanced starting basis, which is then solved to optimality by the Simplex method.

The representation based on the splitting variable approach is motivated by its suitability for solution by the Interior Point methodology. The splitting variable representation is also suitable for developing algorithms based on decomposition such as Lagrangian relaxation.

A simple deterministic model is introduced in Section 2. Section 3 provides an introduction to the general SP approach. The multistage scenario model is presented in Section 3, along with three types of recourse models.

In Sections 4 and 5, the compact and variable splitting representations are presented, respectively. Section 4 contains a detailed description of the *crash procedure* for the compact formulation. Computational results are presented in Section 6.

2. A DETERMINISTIC PRODUCTION PLANNING PROBLEM

Consider the model (2.1) given below. In this model it is assumed that in any given time period, whatever cannot be produced in-house can be produced at a vendor. Production is always completed in the period in which it begins. In each time period, the demand is met through in-house production, inventory and/or vendor production. Our objective is to minimize the total cost of inventory holding, in-house production, and vector supply. (Escudero and Kamesam [13] consider multilevel production planning problems in a manufacturing network). The following notation is used

Sets

J = Set of products $j=1,2,...,|J|$.

T = Set of time periods $t=1,2,...,|T|$.

R = Set of (nontransferable) resources $r = 1,2,...,|R|$, such as machine time, labor, energy, etc.

Deterministic data

h_j = Inventory holding cost per unit of product j in period t

c_{jt} = Per unit cost of in-house production for product j in period t.

p_{jt} = Per unit cost of supply from the second source for product j in period t.

a_{jt} = Amount of resource r that is needed for producing one unit of product j.

k_{rt} = Available amount of resource r in period t.

I_{j0} = Initial inventory of product j.

Variables

x_{jt} = Production volume of product j in period t.

I_{jt} = Inventory of product j at the end of period t.

y_{jt} = Amount of product j obtained from the second source in period t.

Demand data

d_{jt} = Demand for product j in period t.

$$\min \quad \sum_{j\in J, t\in T} h_{jt}I_{jy} + \sum_{j\in J, t\in T} c_{jt}x_{jy} + \sum_{j\in J, t\in T} p_{jt}y_{jy}$$

s.t.   (1) $I_{jt-1} - I_{jt} + x_{jt} + y_{jt} = d_{jt}$ $\forall j \in J,\ t \in T$

     (2) $\displaystyle\sum_{j\in J} a_{rj}x_{jt} \leq k_{rt}$ $\forall r \in R,\ t \in T$ (2.1)

     (3) $x_{jt},\ I_{jt}, y_{jt} \geq 0$ $\forall j \in J,\ t \in T$

If d_{jt} is deterministic, (2.1) can be solved by standard LP software.

If the demand is random, however, we need techniques that can account for the stochastic aspects of the problem. Bitran and Yanesse [4] consider a production planning problem with stochastic demand similar to (2.1). They characterize demand by standard probability distributions and propose deterministic approximations to a *nonsequential decision model* and compute bounds on the optimal value. Bitran and Sarkar [5] extend the results of Bitran and Yanesse [4] to the sequential decision model and show how to compute *a priori* bounds on the optimal value. Sequential decision problems are usually more difficult to solve but they yield better solutions.

In this work we propose a *sequential decision model* that allows for possible recourse actions in the future and develop algorithms that can solve large real world applications. We employ the techniques of stochastic programming as explained in the following sections.

3. NONANTICIPATIVE POLICIES FOR STOCHASTIC PROBLEMS

In multiperiod decision problems the choice between a sequential and a nonsequential model depends on many factors such as i) the severity or importance of the decisions being made in the first stage, ii) whether a rolling horizon approach is used, iii) once a decision is made, what recourse actions (decisions) are possible in future stages. In general, if the first stage decisions have strong consequences, it is advisable to use a sequential decision model. On the other hand, if the first stage decisions can be easily modified in later stages, then a nonsequential decision model can be used on a rolling horizon basis. Sequential decision models are, in general, much harder to solve. As will become evident in the remainder of this section, it is sometimes necessary to construct models in which the decisions are sequential with respect to one set of variables, but nonsequential with respect to others.

Section 3.2 presents three different models with varying degrees of recourse decisions. In each case, we would like to ensure that the model(s) and solution(s) have the following desirable features.

1. the decisions made in each period take into account all future uncertainty.
2. the decisions recommended by the model are consistent with information availability.
3. the model allows for recourse actions in future and optimize with respect to them.

In our test problems, the scenarios specify product demands in each time period. Let S be a finite set of scenarios. We thus have available the demand for product j in time period t under scenario s, say d^s_{jt}, and the weight (probability) associated with scenario s, say w_s, $s \in S$.

The weights represent the likelihood that the decision maker (modeler) associates with each scenario.

3.1. SCENARIO AGGREGATION AND NONANTICIPATIVE POLICIES

Suppose that for each scenario $s \in S$, the table of individual scenario solutions is available (note that each of these problems can be solved by (2.1). Let x^s_{jt} denote the optimal in-house production of product j in time period t under scenario s. Let y^s_{jt} and I^s_{jt} be similarly defined. Condition 2 above, known as the *nonanticipativity* condition requires that the individual scenario solutions be aggregated in such a way as to yield a *nonanticipative policy.* Simply stated, the requirements for this type of policy are:

If two different scenarios s and s' are identical up to time period τ on the basis of the information available about them at time period τ, then up to time period τ the decisions $x^s_{j\tau}$, $F^s_{t\tau}$, $y^s_{j\tau}$ and $x^{s'}_{j\tau}$, $I^{s'}_{t\tau}$, $y^{s'}_{j\tau}$ (see [31]).

Let N denote the set of solutions that satisfy the so called *nonanticipativity constraints.* We would like to ensure that the optimal policy satisfies the above condition. Operationally, this is achieved by adding the following constraint to the optimization model (3.1), (3.2) or (3.3).

$x \in N: = \{x^s_{jt} \mid x^s_{jt} = x^{s'}_{jt}$, for $\forall s$ and s' that are identical up to time t, for $\forall t, j\}$

Similarly, we may impose the conditions $y \in N$ and $I \in N$ to ensure nonanticipativity with respect to y and I. See Escudero et al. [13] for a detailed discussion on these conditions and their desirability. Note that all the scenarios are identical for time period $t=1$ at least, so

$x^s_{j1} = x^{s'}_{j1}$, $y^s_{j1} = y^{s'}_{j1}$ and $I^s_{j1} = I^{s'}_{j1}$, for $\forall j \in J$ and $s, s' \in S$

It is important to understand that although it is necessary to ensure that the optimal solution indeed satisfies the non-anticipativity conditions stated above, computationally, this can be achieved in different ways leading to different problem representations and computational algorithms. These issues are addressed in Sections 4 and 5.

3.2. MODELING WITH RECOURSE

The most appealing aspect of stochastic programming models is the idea of recourse, i.e., modeling what future adjustments or options the decision maker may have and optimizing with respect to them.

They differ from each other only with respect to the extent of recourse actions allowed.

Simple recourse

Suppose that the decision maker first sets the optimal policy $\{x^*_{jt}, y^*_{jt}\}$ for the time horizon T, and that these policies cannot be changed in the time horizon T (even as new information becomes available).

In such cases the only recourse action available to the decision maker is to build inventory to hedge against the uncertainty. The LP model corresponding to these assumptions is as follows.

$$\min \quad \sum_{s \in S} w_s \left(\sum_{j \in J, t \in T} h_{jt} I^s_{jt} + \sum_{j \in J, t \in T} c_{jt} x_{jt} + \sum_{j \in J, t \in T} p_{jt} y_{jt} \right)$$

s.t. (4) $I^s_{jt-1} - I^s_{jt} + x_{jt} + y_{jt} = d^s_{jt}$ $\forall j \in J,\ t \in T,\ s \in S$

(5) $\sum_{j \in J} a_{rj} x_{jt} \leq k_{rt}$ $\forall j \in J,\ t \in T$

(3.1)

(6) $x_{jt},\ I^s_{jt},\ y_{jt} \geq 0$ $\forall j \in J,\ t \in T,\ s \in S$

(7) $I \in N$

Note that in this case, the nonanticipativity constraints (7) are trivially satisfied.

Partial recourse

In all the models presented here, x and y are the control (decision) variables and I are the state variables. There are situations where one set of these control variables (i.e., x or y) can be adjusted over time, but not the other. It is rather easy to allow for such flexibility, because we may impose the nonanticipativity constraints only with respect to one set of the control variables. Model (3.2) below allows recourse with respect to the y variables, but not with respect to the x variables

$$\min \quad \sum_{s \in S} w_s \left(\sum_{j \in J, t \in T} h_{jt} I^s_{jt} + \sum_{j \in J, t \in T} c_{jt} x_{jt} + \sum_{j \in J, t \in T} p_{jt} y^s_{jt} \right)$$

s.t. (8) $I^s_{jt-1} - I^s_{jt} + x_{jt} + y^s_{jt} = d^s_{jt}$ $\forall j \in J,\ t \in T,\ s \in S$

(9) $\sum_{j \in J} a_{rj} x_{jt} \leq k_{rt}$ $\forall r \in R,\ t \in T$

(3.2)

(10) $x_{jt},\ I^s_{jt},\ y^s_{jt} \geq 0$ $\forall j \in J,\ t \in T,\ s \in S$

(11) $I \in N,\ y \in N$

Model (3.2) may be useful for other applications as well. For example, suppose that there is no second source of supply and y^s_{jt} represent the demand that would be lost under scenario s. The cost term in the objective function can be used to effectively control the lost demand.

Full recourse

This is the most general case (hence, more flexible than the others), where all the control variables can be adjusted over time, as more information becomes available. So, the nonanticipativity constraints are imposed with respect to all the variables.

$$\min \quad \sum_{s \in S} w_s \left(\sum_{j \in J, t \in T} h_{jt} I^s_{jt} + \sum_{j \in J, t \in T} c_{jt} x^s_{jt} + \sum_{j \in J, t \in T} p_{jt} y^s_{jt} \right)$$

s.t.   (12) $I^s_{jt-1} - I^s_{jt} + x^s_{jt} + y^s_{jt} = d^s_{jt}$ $\forall j \in J,\ t \in T,\ s \in S$

       (13) $\sum_{j \in J} a_{rj} x^s_{jt} \le k_{rt}$ $\forall r \in R,\ t \in T$

                                                                                              (3.3)

       (14) $x^s_{jt},\ I^s_{jt},\ y^s_{jt} \ge 0$ $\forall j \in J,\ t \in T,\ s \in S$

       (15) $I \in N,\ x \in N,\ y \in N$

If $\bar{Z}$ denotes the optimal expected cost of each of these models, then we should expect

$$\bar{Z}\,(3.1) \ge \bar{Z}\,(3.2) \ge \bar{Z}\,(3.3)$$

See Dembo [8], Dembo and King [9] and Mulvey, Vanderbei and Zenios [28] for more robust models, where the second term in the objective function, $\sum_{j \in J, t \in T} p_{jt} y^s_{jt}$ is replaced by alternatives that certainly offer scenario immunization. Their disadvantage is that nonlinear terms are introduced, thus increasing the computational burden.

4. COMPACT REPRESENTATION

Although the models stated thus far are linear, the dimensionality of the problems to be solved is very large even for moderately sized realistic applications. The size of these models grows exponentially with the number of decision stages. The only practical option open is to approximate the multistage decision process by clustering the time periods into a few decision stages, which in our case is three. Note that "stages" do not necessarily refer to time periods, but rather they represent the stages in the decision process. In Figure 1 below, stage 1 includes time period 1, stage 2 includes time period 2 only and finally stage 3 contains all the remaining time periods (3 thru 12) clustered into a single decision stage.

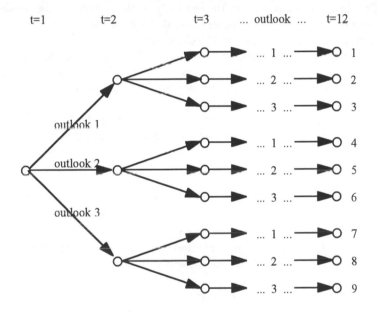

Figure 1. Relation between scenarios and demand outlooks.

All our test cases correspond to a three 'stage decision process approximation for problems with 12 time periods. Even with this approximation, however, some of the problems that we solved have nearly 25,000 constraints and more than 75,000 variables (see Tables 2, 3, and 4 in Section 6). It is clear that we need special algorithms that can solve these problems efficiently. Fortunately, these models have a special structure that can be exploited for computational purposes. In the stochastic programming literature much attention is given to this, leading to specialized algorithms (see [11] for a survey).

In Section 3.2, we presented three different models. All these models have a similar structure and the algorithm that we use to solve them is also the same. For the remainder of this section we will refer to the full recourse model (3.3) only. (In Section 6, computational results are reported on all the models). Similarly, much of the algorithmic description will be for the 3-stage, 9-scenario case represented by Figure 1. This, again, is for notational convenience only, and the principles are applicable more generally.

The problems of interest have the following characteristics:

- The constraint matrix of the overall problem has a nice quasi-staircase structure.
- The different scenarios are distinct only with respect to the product demand. This leads to the observation that within the staircase structure of the constraint matrix, several blocks are repeated.
- There are only a few coupling variables linking the different stages of the problem. Fixing these variables at different stages of the computation leads to a natural decomposition along the stages.

Our computational approach is to solve a single monolithic optimization model by identifying and exploiting its sparsity and special structure.

Sparsity structure

Consider the scenario tree of Figure 1 and the full recourse model (3.3).

Recall that the time periods have been clustered into stages (and, hence, the variables associated with these time periods as well). Figure 2 is a compact representation of the constraint matrix of (3.3), where the constraints (12)-(13) of (3.3) are represented as follows.

Let A_0 be the coefficient submatrix of all the variables X_0 associated with stage 1. Let A_i be the coefficient submatrix of all the variables X_i associated with stage 2 and outlook i for $i=1,2,3$. (Hence scenarios $s=1,2,3$ are associated with A_1, scenarios $s=4,5,6$ are associated with A_2, and scenarios $s=7,8,9$ are associated with A_3). Naturally, there are coupling variables between stage 1 and stage 2 and these are the end of stage 1 inventory variables in the models above. These coupling variables (say, X^{12}) have columns belonging to matrices A_0 and A_i for $i=1,2,3$. Also, let A^j be the coefficient submatrix of the variables X_{ij} associated with stage 3 and demand outlook j for $j=1,2,3$. (Hence, scenarios $s=3(i-1)+j$ are associated with A^j for $i,j=1,2,3$).

Note that if the scenarios differ from each other only with respect to the demand, which indeed is the case with all our test problems, we further have

$$A_1 = A_2 = A_3 \quad \text{and} \quad A^1 = A^2 = A^3$$

The coupling variables between stage 2 and stage 3 are the end of stage 2 inventory variables. Their columns belong to matrix A_i and the matrices A^j for $i,j=1,2,3$. Let these coupling variables be X_i^{23} and let b_0, b_i and b^j be the RHS vectors for the constraints related to stage 1, stage 2 and outlook i, and stage 3 and outlook j for $i,j=1,2,3$, respectively. The following table summarizes the above notation.

Stage	Periods	Variables	Coeff. matrices	RHS
1	1	$x_{j1}^s, y_{j1}^s, \Gamma_{j1}\ \forall j,\ \forall s$	A_0	b_0
2	2	$x_{j2}^s, y_{j2}^s, \Gamma_{j2}\ \forall j,\ \forall s$	A_1, A_2, A_3	b_1, b_2, b_3
3	3,4,...,12	$x_{jt}^s, y_{jt}^s, \Gamma_{jt}\ \forall j,\ \forall s,\ t=3,4,...,12$	A^1, A^2, A^3	b^1, b^2, b^3

4.1. COMPUTATIONAL ASPECTS

Consider the model (3.3). We will partition the coupling and noncoupling variables and, then, the coefficient submatrices accordingly. Recall that the coupling variables between X_0 and X_i, $i=1,2,3$ are X^{12}, and the coupling variables between X_i and X_{ij}, $i, j=1,2,3$ are X_i^{23}. Let

$$\begin{aligned}
X_0 &= (Y_0,\ X^{12}) \\
X_i &= (\quad X^{12},\ Y_i, X_i^{23}) \\
X_{ij} &= (\qquad\qquad X_i^{23},\ Y_{ij})
\end{aligned}$$

and

$$\begin{aligned}
A_0 &= (D_0, D_0^{12}) \\
A_i &= (\quad D_i^{12}, D_i, \qquad D_{i0}^{23}) \\
A_{ij} &= (\qquad\qquad D_j^{23},\ D^j)
\end{aligned}$$

The objective function coefficients are also partitioned in a similar way.

Let $(h_0,\ h_0^{12})$, $(h_0^{12}, h_i, h_{i0}^{23})$ and (h_{i0}^{23}, h^j) be the objective function coefficients of the variables X_0, X_i and X_{ij}, respectively. Note that the h-coefficients include the weights w_s associated with the scenarios. In our test cases,

$$h_1 = h_2 = h_3, \quad h_{10}^{23} = h_{20}^{23} = h_{30}^{23} \quad \text{and} \quad h^1 = h^2 = h^3$$

The compact representation of model (3.3) in terms of the partitioned variables and coefficient submatrices is as follows.

$$Z = \min h_0 Y_0 + h_0^{12} X^{12} + \sum_{i=1,2,3}\left\{h_i Y_i + h_{i0}^{23} X_i^{23}\right\} + \sum_{i,j=1,2,3} h^j Y_{ij}$$

$$
\begin{aligned}
\text{s.t.} \quad & D_0 Y_0 + D_0^{12} X^{12} && \leq b_0 \\
& D_i^{12} X^{12} + D_i Y_i + D_{i0}^{23} X_i^{23} && \leq b_i \; \forall i \\
& \qquad\qquad D_j^{23} X_i^{23} + D^j Y_{ij} && \leq b^j \; \forall i, j \\
& Y_0, \quad X^{12}, \quad Y_i, \quad X_i^{23}, \quad Y_{ij} && \geq b^j \; \forall i, j
\end{aligned}
\tag{4.1}
$$

The special structure of the above representation is also shown in Figure 2. The computational procedures described in the next section take advantage of this structure.

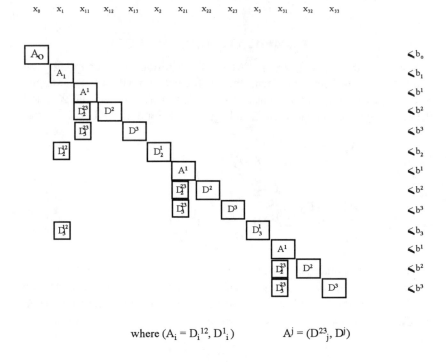

where $(A_i = D_i{}^{12}, D_i^1)$ $A^j = (D^{23}{}_j, D^j)$

Figure 2. Staircase structure of the compact representation.

4.2. SUBPROBLEMS

The crash procedure of the next section solves several LP subproblems of (3.3), as defined below. See also Figure 2.

Stage 1

$$P_0: \quad \min \{c_0 X_0 \mid A_0 X_0 \leq b_0, \ X_0 \geq 0\}$$
$$(4.2)$$

where $c_0=(h_0, h_0^{12})$ is the appropriate vector of the objective function coefficients (products' unit inventory holding cost, production cost, and vendor supply cost) and X_0 is the vector of the unknowns (I, x, y variables) associated with stage 1.

Stage 2

$$P_i: \quad \min \{c_i X_i \mid A_i X_i \leq b_i, \ X_i \geq 0\} \qquad (4.3)$$

where $c_i=(h_0^{12}, h_i, h_{i0}^{23})$ is the vector of objective function coefficients and X_i is the vector of the unknowns associated with the demand outlook I for the time periods in stage 2 for $i=1,2,3$.

Stage 3

$$P_{ij}: \quad \min \{c_i^j X_{ij} \mid A^j X_{ij} \leq b^j, \ X_{ij} \geq 0\} \qquad (4.4)$$

where $c_i^j=(h_{i0}^{23}, h^j)$ is the vector of the objective function coefficients associated with the demand outlook j for the time periods in stage 3 for $i,j=1,2,3$.

4.3. CRASH PROCEDURE

The purpose of this crash procedure is to generate an advanced initial solution for solving (4.1) by the Simplex method. Our procedure for obtaining a hot start (probably primal infeasible, but hopefully dual feasible) is based on decomposing (4.1) into the subproblems P_0 (4.2), P_i (4.3), and P_{ij} (4.4) for all i,j. The subproblems related to earlier stages are solved first. While solving any of the subproblems (4.2), (4.3), or (4.4) for a given stage, the coupling variables between two stages are always fixed to the optimal solution of the earlier stage. The following pseudo code describes the algorithm.

- **begin crash**

1. Solve P_0 and let $\bar{X}_0 = (\bar{Y}_0, \ \bar{X}^{12})$ be the optimal solution.
2. Fix the values of the coupling variables between stages 1 and 2. i.e., set $X^{12} \leftarrow \bar{X}^{12}$.
3. Solve the subproblems P_1, P_2, P_3. Note that because of step 2, P_1, P_2, P_3 can be solved independently. Let $\bar{X}_i = (\bar{X}^{12}, \ \bar{Y}_i, \ \bar{X}_i^{23})$, $i = 1,2,3$ be the optimal solution.
4. Fix the values of the coupling variables between stages 2 and 3. i.e., set $X_i^{23} \leftarrow \bar{X}_i^{23}$, $i = 1,2,3$.
5. Solve the stream of sub problems $\{P_{1j}\}$, $\{P_{2j}\}$ and $\{P_{3j}\}$ for $j = 1,2,3$. Step 4 enables us to solve these subproblems independently. Let $\bar{X}_{ij} = (\bar{X}_i^{23}, \ \bar{Y}_{ij})$ be the optimal solution.
6. $(\bar{X}_0, \ \bar{X}_i, \ \bar{X}_{ij})$ is the initial solution to (4.1).

- **end crash**

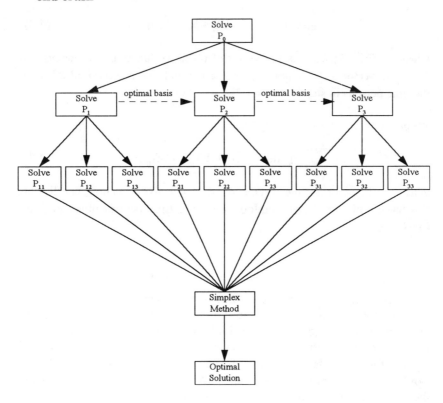

Figure 3. The Crash Procedure.

When the scenarios differ only with respect to the RHS, we have

$$A_1 = A_2 = A_3 \quad\text{and}\quad A^1 = A^2 = A^3,$$

$$h_1 = h_2 = h_3, \quad h_{10}{}^{23} = h_{20}{}^{23} = h_{30}{}^{23} \quad\text{and}\quad h^1 = h^2 = h^3$$

Hence, the computational efficiency of steps 3 and 5 of the crash procedure can be considerably improved. For example, the optimal basis of P_1 can be used as the initial basis for solving P_2. The optimal basis of P_1 is dual feasible for P_2. Similarly, the optimal basis of P_2 can be initial dual feasible basis for P_3.

Similar opportunity exists in solving the subproblems $\{P_{1j}\}$, $\{P_{2j}\}$ and $\{P_{3j}\}$ for $j=1,2,3$ (Because $A^1 A^2 = A^3$ and $h^1 = h^2 = h^3$). The optimal basis of P_{ij-1} can be the initial dual feasible basis for P_{ij} for $j=2,3$. Also, the optimal basis of P_{11} and P_{21} can be the initial basis for P_{21} and P_{31}, respectively.

The generalization of the above procedure to problems with more than 3 stages is straightforward. Varying degrees of parallelism can be employed in solving the above models. It is an open problem to find the appropriate trade-off between parallelism and exploitation of the information provided by the solution of the subproblems solved thus far. Figure 3 summarizes the crash algorithm. The broken lines in Figure 3 show the flow of information between subproblems that will make the computation efficient, but will limit the extent of parallelism. (See also Dantzig and Glynn [7] and Nielsen and Zenios [30]).

5. SPLITTING VARIABLE REPRESENTATION

Representation (4.1), albeit compact, has a sparsity structure that is not suited for Interior Point (IP) methods. (Dense columns in the constraint matrix tend to affect the performance of the IP methods, because they usually cause fill-in and affect the solution of a system of linear equations that has to be solved at every iteration of the IP methods). Figure 4 gives an equivalent representation of (3.3), where the coupling variables X^{12} and X_i^{23} are split into

$$X_0{}^{12} = X_i{}^{12}, \; i=1,2,3 \quad\text{and}\quad X_{i0}{}^{23} = X_{ij}{}^{23}, \; ij = 1,2,3$$

respectively. So, instead of coupling variables, we have the so called coupling constraints with splitting variables between blocks that belong to consecutive stages. The new representation of model (3.3) is as follows. (Joernsten, Nasberg and Smeds [22] and Guignard and Kim [18] use a similar approach in a different context).

$$Z = \min\ h_0 Y_0 + h_0^{12} X_0^{12} + \sum_{i=1,2,3} \left\{ h_i Y_i + h_{i0}^{23} + X_{ij}^{23} \right\} + \sum_{i,j=1,2,3} h^j Y_{ij}$$

$$\text{s.t. } D_0 Y_0 + D_0^{12} X_0^{12} \qquad\qquad\qquad\qquad\qquad\qquad\qquad\qquad \leq b_0$$

$$X_{i-1}^{12} - X_i^{12} \qquad\qquad\qquad\qquad\qquad = 0 \qquad \forall i$$

$$D_i^{12} X_i^{12} + D_i Y_i + D_{i0}^{23} X_{i0}^{23} \qquad\qquad\qquad \leq b_i \qquad \forall i$$

$$(5.1)$$

$$X_{ij-1}^{23} - X_{ij}^{23} \qquad\qquad = 0 \qquad \forall i, j$$

$$D_j^{23} X_{ij}^{23} + D' Y_{ij} \quad \leq b^j \qquad \forall i, j$$

$$Y_0, \quad X_0^{12}, \ X_i^{12}, \qquad Y_i, \qquad X_{i0}^{23}, \quad X_{ij}^{23}, \quad Y_{ij} \ \geq 0$$

where the objective function coefficients and the D constraint matrices are as in representation (4.1). The unknowns are:

$$X_0 = (Y_0, X_0^{12})$$
$$X_i = (X_i^{12}, Y_i, X_{i0}^{23}) \text{ and}$$
$$X_{ij} = (X_{ij}^{23}, Y_{ij})$$

This representation increases the model's dimensions. Lusting, Mulvey and Carpenter [24] explored the advantages of using splitting variable representations for solving deterministic equivalent models to SP models by using IP algorithms.

Although representations (5.1) do not help the Simplex method, they drastically reduce the computational effort while using IP methods because the sparsity structure of the constraint matrix is more suitable for IP methods. Tables 2, 3, and 4 report the computational results based on this strategy.

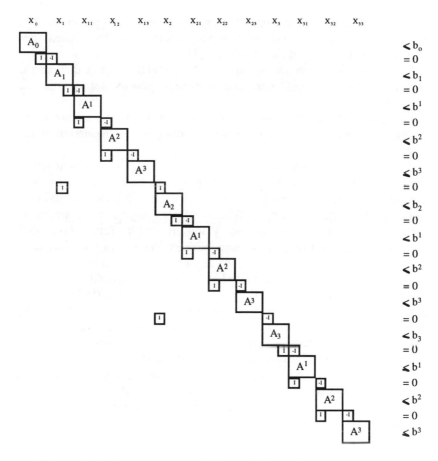

Figure 4. Staircase structure of the splitting variable representation.

Decomposition Methods

Given the sparsity structure of the multistage stochastic programming problems, it is not surprising that many authors studied decomposition methods for solving these problems. Although our approach is not based on decomposition, a brief review of these methods and references is given in this section. Most of these decomposition methods are based on Benders decomposition or (Augmented) Lagrangian dualization of the coupling constraints.

Algorithms based on Benders decomposition take advantage of the dual block angular form of a structure close to (5.1). Implementations described in Birge [2, 3] and Gassmann [17] are based on the L-shaped method introduced by Van Slyke and Wets [38]. These implementations solve subproblems similar to the subproblems solved by our crash procedure. The Lagrange multipliers are updated by solving a relaxed master problem. See also Alvarez et al. [1] and Escudero et al. [14].

Algorithms based on the Augmented Lagrangian method dualize the coupling constraints and introduce nonlinear penalty terms into the objective function.

Dempster [10] in the last section gives a general framework for algorithms based on Lagrangian relaxation, along with a discussion on the problem of dual ascent.

To avoid the difficulties associated with introducing the nonlinear terms, Rockafellar and Wets [31] and Wets [34] proposed the *progressive hedging* algorithm, wherein decomposition is achieved by alternatively fixing a subset of the variables. An application of this algorithm to stochastic networks is given in Mulvey and Vladimirou [29], Rusczcynski [32] proposed a simplicial based procedure. Nielsen and Zenios [30] developed an extension of the *row action* algorithm to stochastic network optimization. Mulvey and Rusczcynski [27] and Rusczcynski [32], on the other hand, developed a diagonal quadratic approximation (DQA) method to deal with the nonlinearities. Alvarez et al. [1] and Escudero et al. [14] developed an algorithm to approximate quadratic nonseparable terms that appear in the Augmented Lagrangian approach by replacing them with (first order approximation) quadratic separable terms.

The Lagrangian decomposition (based on the relaxation of the coupling constraints) solves the same subproblems as in the Benders Decomposition. They differ in the way in which the Lagrange multipliers of the coupling constraints are estimated (updated).

6. COMPUTATIONAL RESULTS

We have developed a software prototype (written in IBM VS FORTRAN V2.0) to test the various representations of the stochastic models. Although the emphasis of this paper is on solving a production planning problem, the implementation is general enough to adapt to other problems.

Scenario creation and test cases

The test problems presented in this section came from a real world setting and all the data describing the products, costs, and capacity utilization came from an actual factory. Faced with the prospect of introducing several new products (to be manufactured in the same factory), the problem on hand was to allocate the different products to in-house and vendor manufacturing lines.

In such cases, it is a fairly common practice within the business world to create several *outlooks* on the demand and then analyze the implications — at least intuitively — of alternative decisions. The demand scenario creation in our test cases is patterned after this (see Figure 1). (Dantzig and Glynn [7], Infanger [20, 21], present an interesting approach of scenario creation for energy systems resource planning problems). We set $w_i=1/|S|$ for all scenarios, i.e., all the scenarios are assumed to be equally likely.

Table 1. Dimensions of the Test Cases.

Test Case	No. of Products	Demand Outlooks	No. of Scenarios
C1	50	3	9
C2	100	3	9
C3	100	5	25

IBM OSL [19] was used as the basic optimization software. All the computational experiments were conducted on an IBM 3090-400E running under VM/XA. The problem sizes are shown in Table 1. The LP dimensions shown in Tables 2, 3, and 4 are for the compact representation. The Simplex method is compared with the Primal–Dual Interior Point method as implemented in OSL (see Forrest and Tomlin [16]). This implementation allows, as an option, the predictor-corrector modification (see Mehrotra [26] and Lusting, Marsten, and Shanno [23]). These results are reported on all the three types of recourse models (3.1), (3.2), and (3.3). The experience gained while testing the different computational approaches is reported in Tables 2, 3, and 4. The following different strategies were tried.

Compact   representation,   Simplex method

SSLV-NO CRASH   Artificial starting basis + Simplex
SSLV-CRASH   Crash procedure + Simplex

Splitting Variable representation,   IP method

BSLV-1   primal-dual IP method
BSLV-2   primal-dual IP method + predictor-corrector modification.

For BSLV-1 and BSLV-2, the *accuracy* reported is the number of significant digits of the optimal value that are accurate. In all the cases, the Interior Point (IP) method was terminated with the same termination criteria.

The IP method, by definition, returns a strictly interior point, whereas the Simplex method will always return an extreme point.

Table 2. Computational experience. Model (3.1): Simple Recourse.

Test Case	C1	C2	C3
number of constraints	4055	8097	25613
number of variables	5074	10148	28000
number of constraints matrix nonzero elements	16641	33282	103500
SSLV-CRASH			
number of iterations	477	819	583
Crash CPU time (secs)	1	5	3
Total CPU time (secs)	4	13	18
SSLV-NO CRASH			
number of iterations	1022	3106	3088
CPU time (secs)	18	25	52
BSLV-1			
number of iterations	35	35	28
accuracy (number of digits)	4	4	5
CPU time (secs)	14	21	37
BSLV-2			
number of iterations	34	34	23
accuracy (number of digits)	4	4	5
CPU time (secs)	10	15	16

Several remarks can be made from the results shown in the tables. In general, the compact representation solved by the *crash* procedure and, then followed by the Simplex method did very well. The performance improvement gained by the *crash* procedure is very satisfying and worth the effort.

The compact representation when solved by the IP method did not do well. The IP method, however, required only a small number (25 to 35) of iterations to solve the splitting variable representation, although the solution was not very accurate. It has been our experience that the predictor-corrector modification (in OSL) improves the performance of the IP method in terms of the CPU time required to reach a termination criteria, but the number of iterations required and the accuracy of the solution are comparable to the IP method without the predictor-corrector modification. In general, the Simplex method performs better on the compact representation and the IP methods perform better on the splitting variable representation.

Based on our experimentation, the computational effort required by the IP method with the predictor-corrector modification for the splitting variable representation (BSLV-2) is comparable with the computational effort required by the Simplex method along with our *crash* procedure, applied to the compact representation. The strategy SSLV-CRASH, however, gives more accurate solutions. Although problem C3 is larger than C1 and C2, SSLV-CRASH ran faster on C3 because the starting solution generated by the crash procedure was an extremely good one.

From the results of the strategy SSLV-NO CRASH, it is clear that even with very advanced Simplex codes, it is still necessary to develop specialized algorithms to solve very large problems with structure.

Table 3. Computational experience. Model (3.2): Partial Recourse.

Test Case	C1	C2	C3
number of constraints	4055	8097	25613
number of variables	8600	17200	52400
number of constraints matrix nonzero elements	28780	57547	180313
SSLV-CRASH			
number of iterations	2265	5039	221
Crash CPU time (secs)	11	29	4
Total CPU time (secs)	29	143	7
SSLV-NO CRASH			
number of iterations	5509	12582	37449
CPU time (secs)	96	528	4674
BSLV-1			
number of iterations	29	34	25
accuracy (number of digits)	4	4	5
CPU time (secs)	19	61	346
BSLV-2			
number of iterations	29	34	25
accuracy (number of digits)	4	5	4
CPU Time (secs)	17	38	239

The *crash* procedure given in Section 4.3 fixes the values of the coupling variables to the optimal values of the subproblem solved earlier. The main motivation for this is that in multistage scenario based optimization where only the RHS changes from scenario to scenario, this strategy leads to subproblems that are identical except for the right hand side (See Figure 3. The subproblems P_{11}, P_{12}, P_{13} are identical but for their RHS).

The solution generated by the crash procedure is, in general, not feasible for (4.1), but the total primal-dual infeasibility is small. The most important feature of the crash solution is that it is a *basic* solution, whereas the solution generated by the IP method is *nonbasic*.

We thus are able to solve the subproblems very efficiently. It is worth noting that in one test case (Table 4, C3), the Simplex method failed to converge starting from an artificial basis or starting from the solution of BSLV-1, where as it converged starting from the solution of the crash procedure or the solution of BSLV-2.

Table 4. Computational experience. Model (3.3): Full Recourse.

Test Case	C1	C2	C3
number of constraints	4137	8179	25857
number of variables	12126	24252	76800
number of constraints matrix nonzero elements	32388	64252	204957
SSLV-CRASH			
number of iterations	3587	6849	4378
Crash CPU time (secs)	7	28	37
Total CPU time (secs)	84	259	80
SSLV-NO CRASH			
number of iterations	7704	16866	>51194
CPU time (secs)	397	1666	>7 hrs
			elapsed
BSLV-1			
number of iterations	28	27	20
accuracy (number of digits)	5	5	5
CPU time (secs)	21	47	254
BSLV-2			
number of iterations	27	27	20
accuracy (number of digits)	5	5	5
CPU time (secs)	14	31	177

Passing the solution of the IP method to the Simplex method to improve the accuracy of the solution (as well as to obtain an optimal basis) is an obvious possibility. This proved to be computationally very expensive for a long time. (Our early results were in line with this experience). Recent implementations based on an algorithm given in Meggido [25] seem to have overcome this difficulty to a great extent. So, in future this may be an effective alternative.

7. CONCLUSIONS

We have presented a comprehensive approach to modeling the aggregate production planning problem under stochastic nonstationary demand, capacity constraints, and dual sourcing. As products and manufacturing become more and more standardized, manufacturing needs are being increasingly met through capacity at other manufactures/vendors rather than by building in-house capacities to meet all contingencies. The dual sourcing aspect of our models is very important for this reason.

All the models presented in this paper are prescriptive in nature. Two equivalent representations are proposed for the deterministic equivalent of the stochastic model. The compact representation approach is more suitable for the Simplex method than for the Interior Point method. Although our test instances resulted in extremely large models, much to our surprise, we were able to solve them with relative ease, mainly due to the effectiveness of our crash procedure to exploit the staircase structure of the models.

On the other hand, although the splitting variable representation results in larger problems, its constraint matrix is suitable for the Interior Point methods. Even though the pair compact representation Simplex methodology is the winner in our experimentation, the results for the alternative approaches are good enough given the model's dimensions.

In this paper we have reported the results from a large computational study. A computational study comparing different algorithmic approaches with and without decomposition will add to the understanding of the relative merits of these approaches. We are also experimenting with algorithms based on Lagrangian decomposition of the splitting variable representation, but they are unlikely to improve the computational times of the test problems in this paper. For much larger problems, or when the model is nonlinear, decomposition approaches are expected to play an important role. (See Alvarez et al. [1] and Escudero et al. [14].) Also, decomposition may be the only viable approach for large stochastic mixed integer programming (MIP) problems.

Stochastic programming models with recourse offer a promising direction for solving real world planning problems. Although the need to account for the uncertainties is well recognized, much of the practice to date is still limited to using deterministic models. One of the main reasons for this is that until very recently it was not possible to solve large stochastic models. The explosive growth in computing power, massively parallel computer architectures, and new efficient algorithms should go a long way in bringing the stochastic programming methodology to the practitioners table.

REFERENCES

1. **Alvarez M., Escudero L.F., de la Fuente J.L., Garcia C.** and **Prieto F.J.**, Network planning under uncertainty with an application to hydropower generation, TOP 2 (1994) 25-58.
2. **Birge J.B.**, Decomposition and Partitioning Methods for Multistage Stochastic Linear Programs, Operations Research 33 (1985) 989-1007.
3. **Birge J.B.**, An L-Shaped method computer code for multi-stage stochastic linear programs, in: Y. Ermoliev and R.J.-B. Wets (Eds.), Numerical techniques for stochastic optimization, Springer-Verlag, Berlin, (1988) 255-266.,
4. **Bitran G.** and **Yanesse H.**, Deterministic approximations to stochastic production planning problems, Operations Research 32 (1984) 999-1018.
5. **Bitran G.** and **Sarkar S.**, On upper bounds of sequential stochastic production planning problems, European J. of Operational Research 34 (1988) 191-207.
6. **Bjornestad S., Hallfjord A.** and **Joernsten K.O.**, Discrete Optimization under uncertainty: The scenario and policy aggregation technique, Working paper 89/06, Chr. Michelson Institute, Fantogt, Norway, 1989.
7. **Dantzig G.B.** and **Glynn P.W.**, Parallel processors for planning under uncertainty, Annals of Operations Research 22 (1990) 1-21.
8. **Dembo R.S.**, Scenario immunization, in: S.A. Zenios (Ed.) Financial optimization, Cambridge University Press, Cambridge, USA, (1991).
9. **Dembo R.S.** and **King A.J.**, Tracking models and the optimal regret distribution in asset allocation, Research report RC-17150, IBM T.J. Watson Research Center, Yorktown Heights, NY., (1991).
10. **Dempster M.A.H.**, On Stochastic Programming II: Dynamic problems under risk, Stochastics 25 (1988) 15-42.
11. **Ermoliev Y.** and **Wets R.J.B.** (Eds.), Numerical techniques for stochastic optimization , Springer-Verlag, Berlin, (1988) 255-266.
12. **Escudero L.F.** and **Kamesan P.V.**, MRP modelling via scenarios, in: Ciriani and R.C. Leachman (Eds.), Optimization in industry, John Wiley, London, (1993) 101-111.

13. **Escudero L.F., Kamesan P.V., King A.** and **Wets R.J.B**, Production planning via scenario modelling, Annals of Operations Research 43, (1993) 311-355.

14. **Escudero L.F., de la Fuente J.L., Garcia C.** and **Prieto F.J.**, On solving multistage stochastic linear networks, Mathematical Programming (1994), submitted for publication.

15. **Eppen G.D., Martin R.K.** and **Schrage L.**, A scenario approach to capacity planning, Operations Research 37 (1989) 517-527.

16. **Forrest J.J.** and **Tomlin J.A.**, Implementing Interior Point Linear Programming methods in the Optimization Subroutine Library, IBM Systems Journal 31 (1992) 26-38.

17. **Gassmann H.I.**, MSLiP: A computer code for the multistage stochastic linear programming problem, Mathematical Programming 47 (1990) 407-423.

18. **Guignard M.** and **Kim S.**, Lagrangian Decomposition: A model yielding stronger Lagrangian bounds, Mathematical Programming 39 (1982) 215-228.

19. **IBM**, OSL: Optimization Subroutine Library, Guide and Reference, SC23-0519, (1990).

20. **Infanger G.**, Monte Carlo (Importance) Sampling within a Benders decomposition algorithm for stochastic linear programs, Technical report SOL. 89-13, Systems Optimization Laboratory, Stanford University, Stanford, CA., (1989),

21. **Infanger G.**, Planning under uncertainty, Boyd & Fraser Publishing, Danvers, MA, (1994).

22. **Joernsten K.O., Nasberg M.** and **Smeds P.A.**, Variable Splitting: A new Lagrangian relaxation approach to some mathematical programming models, Report LITH-MAT-R-85-04, Linkoping Institute of Technology, Linkoping, Sweden (1985).

23. **Lusting I.J., Marsten R.E.** and **Shanno D.F.**, On implementing Mehrotra's Predictor-Corrector Interior Point Method for Linear Programming, SIAM J. of Optimization, 2 (1992) 435-449.

24. **Lusting I.J., Mulvey J.M.** and **Carpenter T.J.**, Formulating two-stage stochastic programs for interior point methods, Operations Research 39 (1991) 757-769

25. **Meggido N.**, On finding Primal and Dual optimal bases, ORSA J. on Computing 3 (1991) 63-65.

26. **Mehrotra S.**, On the Implementation of a Primal-Dual Interior Point Method, SIAM J. of Optimization 2 (1992) 575-601.

27. **Mulvey J.M.** and **Ruszczynski A.**, A diagonal quadratic approximation method for large scale linear programs, Operations Research Letters 12 (1992) 205-216.

28. **Mulvey J.M., Vanderbei R.J.** and **Zenios S.A.**, Robust Optimization of large-scale systems: General modelling framework and computations, Technical report 91-06-04, The Wharton School, University of Pennsylvania, Philadelphia, PA, (1991).

29. **Mulvey J.M.** and **Vladimirou H.**, Stochastic network optimization models for investment planning, Annals of Operations Research 20 (1989) 187-217.

30. **Nielsen S.S.** and **Zenios S.A.**, A massively parallel algorithm for nonlinear stochastic network problems, Operations Research 41 (1993) 319-337.

31. **Rockafellar R.T.** and **Wets R.J.B.**, Scenario and policy aggregation in optimization under uncertainty, Mathematics of Operations Research 16 (1991) 119-147.

32. **Ruszczynski A.**, Interior Point methods in stochastic programming, Working paper WP-93-8, IIASA, Laxenburg, Austria, (1993).

33. **Van Slyke R.** and **Wets R.J.B.**, L-shaped linear programs with application to optimal control and stochastic programming, SIAM J. of Applied Mathematics 17 (1969) 638-663.

34. **Wets R.J.B.**, The aggregation principle in scenario analysis and stochastic optimization, in: S.W. Wallace (Ed.), Algorithms and model formulations in mathematical programming, Springer-Verlag, Berlin, (1989) 92-113.

PART 3

CASE STUDIES

Chapter 7

DEVELOPMENT OF A DESIGN METHODOLOGY FOR TOOL LOGISTIC SYSTEMS IN FLEXIBLE MANUFACTURING

M. Calderini, M. Cantamessa
Dipartimento di Sistemi di Produzione
e di Economia dell'Azienda
Politecnico di Torino, Italy

Vincenzo Nicolò
Vice President Consortium COSPI
Consultant, Piacenza, Italy

1. AN OVERVIEW OF THE TLS DESIGN PROBLEM

Experience in Flexible Manufacturing Systems (FMSs) has shown that their efficient operation, as well as a satisfactory return on the large investments needed for their implementation, is closely connected to the design quality of both logistic structures and management strategies. Looking at an FMS in action, it is impressive to see the large number of pallets, fixtures, tools and other equipment being moved around the plant by various kinds of transportation devices, such as AGVs, robots, and others. All these actions have to be carefully planned, coordinated, controlled, and monitored, so that the FMS may work at the top of its performance, such a task being remarkable indeed.

A crucial role in the global FMS design and management problem is covered by the "Tool Logistic System" (TLS). Responsibility of this subsystem is to store and transport tools in order to ensure the availability of the necessary tools at the right time to the different machining centers (MCs), based on the production plan assigned to the system and on the tooling requirements associated with each part's process plan. Moreover, the TLS has to handle and manage the flow of worn and broken tools to the reconditioning facility present in the Tool Room. From the functional point of view a TLS may therefore be seen as the interface between the Tool Room and the MC spindles; from an operational point of view, it is also clear that TLS management needs to be closely connected both to the part management software architecture, and to the real-time monitoring system.

As will be discussed later, the common approach to FMS design and management is to look at it under the point of view of parts, assuming other logistic aspects involved in FMS operation (mainly tools and fixtures logistics) as external constraints. As a consequence the TLS is generally oversized by system designers, both for what concerns the number of physical tools available in the plant, and for handling the devices performance. Evidence for this assertion comes not only from industrial practice, but also from all the research papers based on the above mentioned assumption. Such a philosophy clearly simplifies the design task, but implies redundancy costs which can hardly be neglected in the perspective of minimizing FMS investment costs [26]. In industrial practice such redundancy often turns out to be even greater, since TLS design is often performed with rule of thumb methods supported by simulation.

Recent trends in FMSs yield further evidence to the importance of a serious study of the design and management of efficient tool logistic systems [8, 11, 17, 32]. Pressure on tooling systems is being increased by the flexibility requirements being cast on FMSs to reduce lot sizes and, therefore, time available between setups. In addition, the introduction of high-speed machining, seen as a promising innovation for the future [20], is sensibly shortening processing times; this factor also leads to reducing time available between different setups, thus leading to a more intense flow of tools in the plant. Moreover, tools for high-speed machining are expensive, and this constitutes a serious deterrent against using duplication as a practical way to ensure tool availability. As a last consideration, many medium-sized industries consider FMS investments hazardous and not remunerative enough. The need to obtain a higher return in financial terms from FMS operation emphasizes the importance of making such plants work in totally unmanned shifts; for such kind of operation, the availability of efficient and reliable tool management becomes a crucial issue.

Looking at literature it is remarkable to note that among the dozens of papers existing on flexible manufacturing systems, not much may be found on TLS design and its management. Starting from the management point of view, and setting aside the many papers which deal with FMS scheduling without considering the tooling aspect [10, 30], the subject of FMS management with tooling considerations has constituted for many researchers, mainly from the Operations Research field, a challenging problem. The FMS machine loading/tool allocation problem is discussed by a number of authors [2, 3, 12-14, 27, 33-38, 43] while scheduling with tooling considerations is covered by [8, 19, 25, 39, 40, 42]. A certain number of these papers contain approaches which are applicable in industrial environments, and may therefore constitute a "library", though still unorganized, of practical solutions. However, most of these approaches aren't applicable, either because of the computational complexity deriving from the models adopted, or because of their difficulty to relate them to actual manufacturing environments; unrealistic assumptions about the actual technological scenario mostly concern the inflexibility of part programming. It is also noteworthy that most theoretical approaches deal with oversimplified Tool Logistic Systems, in which many approaches of common use in industry are overlooked. For example, no paper deals with multilevel tool storage, and very few with real-time tool sharing among different machining centers [7, 19].

All the papers mentioned deal with tools and TLSs with given constraints and not as possible design variables. On this side, tool requirements planning is, in fact, covered by very few authors [11, 44] while, considering FMS design, the only specific book on the subject [41], does not mention Tool Logistics as a design issue.

Industrial designers, therefore, have a serious problem when having to design a TLS for a new system, or for retrofitting an existing one. Design of TLSs requires the use of well-founded models able to support the basic steps of design, namely the development of system architecture and of management policies. Since such steps must be carried on in close connection with the logistics of parts, these models should be easy to interface with it. Secondly, it is then necessary to choose the system components and decide the actual geometrical layout of the system. Means should be provided so that passing from system architecture to the final design and floorplan be an easy and unambiguous task (possibly implementable in a computer aided design system).

In this paper the problem of TLS design is at first discussed in Section 2, where it is shown that a classification and representation framework for TLSs is required as a basis for design. Section 3 then deals with the proposal of a comprehensive methodology for TLS design, based on the classification framework presented in Section 2. Conclusions and propositions for future research will then be drawn in Section 4.

2. ANALYSIS OF THE PROBLEM

2.1 A CLASSIFICATION FRAMEWORK FOR TLS LAYOUT TYPOLOGIES

Once identifying the issues which give relevance to the Tool Logistics problem, it is necessary to outline the role which is played by TLS design within the task of designing the whole FMS. It will be shown that the attempt to integrate TLS definition in the general FMS design process implies building up a classification framework for such systems. As shown in Figure 1, the design of an FMS can be imagined as composed of four distinct phases, aimed respectively at the design of the management strategies and of the physical logistic systems both for parts and for tools.

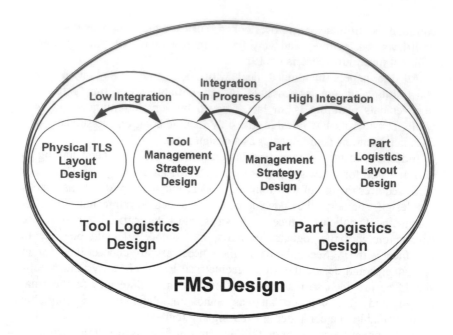

Figure 1. The main phases of the FMS design process.

Obviously, an efficient design process requires these phases be performed as concurrently as possible in order to allow the activation of close and frequent refinement loops; as a matter of fact however, the level of integration among these different design tasks is, at the present state of the art, very different. On one side, the integration between the design of physical part logistic systems and part management strategies is quite consolidated. More efforts are also being placed in increasing the level of integration between part management and tool management strategies. Evidence for this is given by the number of research papers dealing with the joint scheduling of tools and parts. Conversely, very little can be seen concerning the integration among physical tool logistic system design and the other three phases. Reasons for this can be many, almost all originated by the fact that TLS design has always been considered as a subordinate problem to the general FMS design problem and that the tool resource has been typically considered as a constraint for parts management strategies instead of a design variable, thus originating the problems outlined in section 1.

In general it is also possible to ascribe such low integration to the *lack of identity* of layout configurations. The term "identity" means a set of attributes which enables the various TLS layout typologies to play an active role in the design process. These attributes should basically point out features describing the suitability of a TLS layout typology to cope with a given production scenario and with selected tool management strategies. Giving an "identity" to TLS layout types therefore implies the development of a classification scheme based on those attributes which are relevant in terms of management strategies and production scenarios. Such "identity" is to be searched for under a *functional* point of view, since it must tell how the TLS works (which is closely connected to the management problems), rather than under the *physical* point of view (that is, how it looks), even though some relationship between these two points of view undoubtedly exists.

The "identity" allows precisely describing the various types of TLSs available, and such description may be used both for synthesis and analysis problems. Basically two situations can be imagined; on one hand the physical structure of the TLS may have to be designed, thus involving a synthesis problem: desired attributes for the TLS have to be inferred from the design specifications, and a layout configuration is to be selected on the basis of such attributes. The task is then to "describe" the configuration in order to make it "visible" and "utilizable" for other design tasks, in particular for defining tool management strategies. On the other hand, the physical structure of the TLS may be predefined, thus being at first a design constraint for the definition of management strategies and, if necessary, the starting point for an innovation process. In this case the analysis task requires that the given TLS be described in terms of its functionality by starting from its geometrical appearance.

The set of features to be pointed out in order to define TLS layout identity is very wide and could generate an extremely complex classification framework. In order to point out the most relevant aspects of TLS operations, a survey of about sixty FMS of different European builders has been performed. As a result of the survey, it has been found that the TLS, which is the environment where tools are moved and stored in order to perform their processing tasks, may essentially be seen under two main functional aspects: the TLS as a transport device and as a storage device [6]. The classification scheme is therefore built upon this basis.

2.1.1 Functional classification (1). The TLS as a transportation system

A first aspect in the functional classification is connected to the TLS as a means for moving tools. To fully understand the criteria at the base of the classification scheme it is necessary to set aside the usual geometrical definition of "cell" and "line" layouts, which basically distinguishes incident-axis machine configurations and parallel-axis machine configurations. Similarly, the distinction between cell and line has no connection — in principle — with software architecture. In this paper, as also stated in [23, 24], the concept of cell and line refers to the relationship between the machining centers and the transport system configuration. Under a transportation perspective, a formal definition of "cell" and "line" may be given:

Definition 1. a layout has a "line" configuration if each machining center is served by a transport device which can serve all other machining centers located in the plant.

Definition 2. a layout has a "cell" configuration if in the plant it is possible to identify groups of machining centers which are served by independent transport systems (eventually linked together by a higher-level transport device).

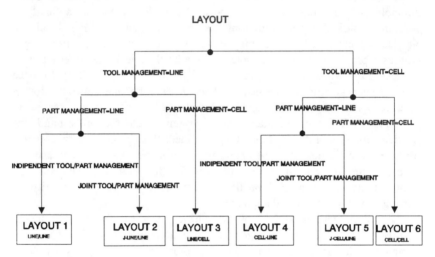

Figure 2. Classification of TLSs as transportation systems.

This criterion may be applied both to part and tool transport, and allows identifying four different configurations generated by the four possible combinations, as illustrated in Figure 2.

The same figure shows that the classification tree may be further branched in order to outline the existence of joint part/tool transportation systems. As will be specified further on, this configuration bears many implications for what concerns production management. Such a distinction has not been applied to one of the four identified cases, since up to now it has been found to be of no practical relevance. Six prototypical configurations have therefore been identified.

2.1.2. Functional classification (2).
The TLS as a hierarchical storage system

The second part of functional modeling is related to the TLS as a means for storing tools. The TLS may be described as a multilevel structure, where tool flows can be imagined descending from the tool room to the spindle through different buffering stages, as illustrated in Figure 3.

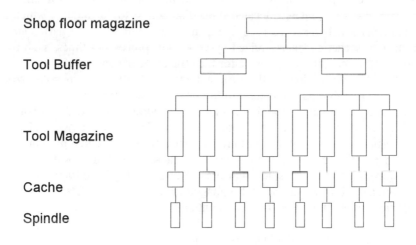

Figure 3. Classification of TLSs as storage systems.

At the upper level of such hierarchy it is possible to find the *shop floor magazine*, which is a tool storage device common to all the machining centers in the FMS; it may basically be considered as an online tool room (e.g., a tool room without refreshing and tool configuration functions). The shop floor's capacity should be able to store all of the tools present in the plant.

At a lower level one can find *tool buffers* which are decentralized storages dedicated to a group of machining centers. The name itself gives the idea of the function of this storage level where tools are kept for two main reasons. First, it represents a standby position for tools which are to be moved to an upper or to a lower storage level; it also constitutes a storage position for tools which, for reasons which will be explained further on, are to be shared between Machining Centers. In other cases, when the number of tools to be directly stored in Machining Centers is very high, instead of installing very large tool magazine structures on board Machining Centers (thus affecting both economics and flexibility of the system), it might be more advisable to decentralize some storage capacity on the tool buffer level.

Always descending the illustrated structure are *tool magazines*, which are storage devices oriented to serving a single Machining Center. For this reason they are also called "on-board" magazines, even though they are often located aside the Machining Centers.

The tool magazine and the spindle are generally connected directly by an automatic tool changer. However, it is possible to identify an additional storage level, represented by an optional *cache magazine*. This is a small and very fast storage device located immediately next to the spindle, which stores tools which are to be used again in a very short time. In particular production scenarios, characterized by very short processing times, such a device can have a strong relevance for the efficient management of tools.

After this description of the different storage levels, it is possible to outline at least four different configurations of the tool storage architecture, depending on the number and typology of the storages provided in plant. Tool Magazines are always considered part of TLS architecture, while shop floor magazines and tool buffers may alternatively be considered present or not. In order to not burden the classification scheme, Cache Magazines aren't taken into account, since their presence has mainly local implications and doesn't affect the general tool management problem. For these reasons, realistic configurations taken into account are:

A) tool room-shop floor-tool buffer-tool magazine-spindle
B) tool room-shop floor-tool magazine-spindle
C) tool room-tool buffer-tool magazine-spindle
D) tool room-tool magazine-spindle.

2.1.3 Integration of the transport and storage classifications: the final classification scheme

After having defined two disjointed classification schemes for transport and storage functions, it is now time to integrate them into a single final classification framework. The result, illustrated in Figure 4, derives from the combination of the six different transport layouts and the four storage architectures previously identified. The survey proved that all the twenty-four configurations are of practical relevance.

Figure 4 may be used as a tool for analyzing an existing TLS as well as for carrying out a design task. In the former it is sufficient to associate the existing layout to one of the twenty-four models, and the description of the functional behavior of such TLS from both the transport and the storage points of view may immediately be found. In the latter task, the designer may, based on the desired attributes, go over the two decisional trees until he arrives at one of the models. At this point, TLS layout design simply implies instantiating the "standard model" and choosing the appropriate components.

2.2 RELATION WITH THE TECHNOLOGICAL AND PRODUCTION SCENARIO (2)

The classification framework previously presented yields a comprehensive set of twenty-eight different layout architectures, whose graphical representation is not possible to give here, due to lack of space. Upon the basis of the classification framework, it has been possible to develop a rule-based selection procedure able to aid the designer in picking up the most suitable TLS with respect to the production scenario the plant has to work in. The principles on which such procedure has been built will now be discussed.

Since the classification scheme has separately dealt with the two main functions of tool logistic systems (namely transport and storage), the discussion will be carried on accordingly, leaving to the reader the task of combining the two as shown by Figure 4.

2.2.1 Criteria relative to the TLS as a transportation system

As far as tool transport aspects are concerned, we won't develop specific considerations about the performance of single physical components, giving more emphasis to the management strategies deriving from the selected transport logistics.

At first it has to be stated that given recent practice in scheduling [28], more simple and efficient part and tool management may be performed with the assumption that either parts logistics or tools logistics is connected to the other one. In other words, two basic alternatives are taken into account: to move tools minimizing part transfers or to move parts minimizing tool transfers. Since the main feature of a cell configuration is to have a local transport system, it has to be considered as a typically static solution. On the contrary, due to its structure, line configuration is a more dynamic and flexible solution. In general, mixed solutions (managing the "mobile" objects by line and the "fixed" ones by cells) may be preferred rather than homogeneous solutions. For these reasons, and as a general statement, layouts C and D are to be considered as target configurations, since they are highly adaptable to efficient management strategies.

Moreover, it is well known that industrial practice indicates cell configuration as the best one if related to the classic line configuration, since it implies a much easier management software implementation. In particular, as part logistics is often more difficult to be managed than tool logistics, a very good configuration seems to be layout C. In order to implement such a solution, it is nevertheless necessary to have cells composed by extremely flexible MCs, since they are required to start and complete the processing of a part within a single cell. In case parts should require very peculiar operations, and consequently be moved around the plant, it could be better to adopt a solution such as layout D, where tool management is fairly simplified. It has to be noted that such a solution requires a higher number of physical tools (i.e. duplicates per tool type) present in plant; this could become not affordable in case expensive tools for particular processing tasks are to be used.

Finally, a cell-cell solution (layout F), which has to be considered the simplest one in terms of management strategies implementation, although scarcely flexible under the aspect of part routing modifications, might represent the result of applying Group Technology concepts [18, 21], which should allow the TLS designer to assign to the same MCs, parts with similar tool sets [5].

A last consideration is to be reserved to layout architectures B and E, whose peculiarity is to have a joint parts and tools logistics. This solution is considered realistic, as demonstrated by the wide number of industrial applications found in the survey [14]. Having parts and tools managing with the same physical structure allows significant reduction of investment costs, but it has to be noted that software required to manage such solution is definitely sophisticated. The applications of joint logistic solutions are therefore to be limited to the cases which present a low complexity of the tool and part management problem, especially in terms of object transfer frequency. Moreover, such solutions are practically realized with RGVs or AGVs, thus implying the execution of tool transport by batches. Translated into production data, this mainly means large part batch sizes, low tool wear, and long cycle times. Infrequent adaptability to complex management systems also makes Layouts B and E not suitable for plants including a large number of Mcs.

2.2.2 Criteria relative to the TLS as a storage system

Regarding the selection of the four prototypical configurations pointed out within the tool storage classification framework, it is immediate that the four storage architectures are discriminated by the presence or absence of the tool buffer and the shop floor magazine levels. It is therefore necessary to understand the role played by the two in a multilevel storage architecture for tools.

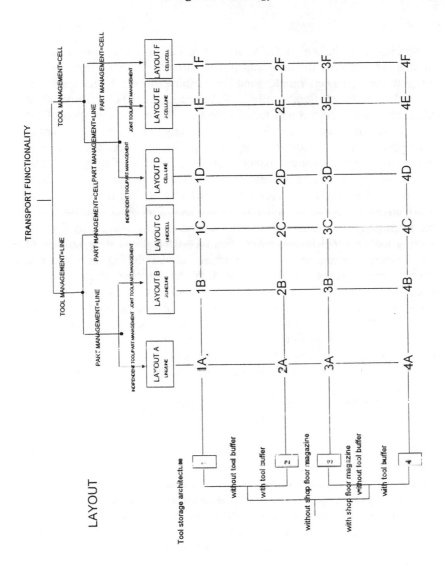

Figure 4. The final TLS classification scheme.

Starting from the tool buffer level, its two functions are: spreading tool storage capacity over the plant and fractionating tool storage capacity into different levels. To better understand the meaning of the former function, it has to be noted that, traditionally, tool storage capacity is concentrated on MCs, bringing in such a way to an almost totally local distribution of the tooling resource and, therefore, making any kind of tool sharing management strategy difficult to be applied (except for scenarios characterized by a limited number of tool types). Tool sharing (or "tool migration", as sometimes defined) strategies have a number of advantages which have been recently analyzed in literature [1, 7, 19], the most evident of which are the drastic reduction of the number of physical tools needed in the plant and of the number of part transfers among machining centers. Storing tools on tool buffers makes them available for a wide number of MCs present in plant; therefore, tool buffers allow reducing the number of physical tools present in plant. Regarding the function of fractionating storage over different levels, tool buffers decrease the required storage capacity of Tool Magazines, since part of the tools are moved to the upper level. Small Tool Magazines are an advantage, since they have lighter structures and faster operative times; in addition, the cost per storage slot is much higher for Tool Magazines than for Buffers.

It is now necessary to clarify which features of production scenarios may suggest TLS designers to prefer storage architectures with tool buffers, especially for deciding on the tool sharing option. A parameter to be taken into account is the "tool variance" with respect to MCs and part programs: in other words, it has to be understood whether tool sets assigned to different MCs and part programs are very similar or extremely differentiated. In the case of a production scenario where most tool types are used by all MCs present in plant, a trade-off analysis has to be performed: such a situation may be coped with by overdimensioning the number of physical tools present in plant and, therefore, the global storage capacity. On the opposite, this can be avoided by storing tools in an intermediate position so to make them available for more MCs; however, this second option becomes more difficult to administer, since tool management is now to be performed at the general plant level and not only locally on single MCs.

Configurations 2 and 4 are therefore suitable to production scenarios involving the intense tool traffic due to tool sharing. Moreover, the use of a batch tool transport system, which implies high peaks in tool flows, stresses the importance of having buffering stages along the journey of tools from the tool room to the spindle.

Focusing attention on specific technological aspects, the case of very short processing times per operation, and high number of different tools associated to each part type also requires an articulated multilevel storage capacity. Tool Magazine overloading, which represents the most common industrial practice until now, might not be feasible in such cases, since a too large and heavy Tool Magazine can prevent the Tool Magazine from performing sufficiently quick tool exchanges. At the extreme, tool search and exchange time might result in longer cutting time on the spindle. If this happens, one of the main concepts of FMS philosophy is to realize MC setups in masked time could be lost. To avoid this, Tool Magazines structures should be made smaller and faster by moving some of their storage capacity to an upper level of the storage architecture.

On the opposite side, in presence of large lots and high tool wear, tool changes due to tool wear will be more frequent than tool changes due to production change. This implies that tools presumably begin and end their lives on a single MC; consequently, tool flows are directed from the Tool Room to MCs, with no need for tool sharing among different MCs. This aspect becomes even more relevant when the MCs in the plant are typically dedicated working units, which require specific tools. Having little need for tool sharing suggests providing batch or single tool deliveries directly connecting the Tool Room and Tool Magazines, and layout 1 emerges as the most suitable to cope with this kind of production scenario.

Let us now consider the same production scenario indicated for layout 1, but with the necessity of performing totally unmanned shifts. In most cases it won't be possible to store all the tools required to perform an eight hour production shift on Tool Magazines. For this reason a shop floor magazine (layout 3) may help in adding storage capacity to the plant; such a device, able to store a number of tools sufficient for an eight hour production shift, is logically equivalent to a fully automated Tool Room. Another interesting case to which layout 3 could be applied is the case of a Tool Room dedicated to several FMS lines, where the shop floor magazine may have the role of buffering Tool Room activities and FMS operations.

While the meaning of the upper storage levels is quite evident, the idea of providing an additional capacity between the Tool Magazine and the spindle seems to be unfamiliar to the majority of MCs builders. However, in case of very short cutting times and frequent tool exchanges, it is useful to provide such a small storage (5–10 tools), located close to the spindle, to keep tools of most frequent use. The addition of this storage capacity, named Cache Magazine in the classification scheme, represents a little variant to each of the four prototypical architectures identified within the tool storage classification framework.

3. A METHODOLOGICAL APPROACH TO THE TLS DESIGN PROBLEM

3.1 REQUIREMENTS CAST ON THE METHODOLOGY

The classification scheme discussed above is the conceptual basis upon which a methodology for TLS design has been developed. Such methodology has then brought to the specification of a Decision Support System for the design of TLSs which is currently being implemented.

In order to explain the methodology, it is necessary to point out some considerations regarding the development of design methodologies for complex systems. It is well known that there exists a wide body of literature for design theories and methodologies, but their application is difficult in systems with high complexity, such as manufacturing systems; the term "complexity" here means both the "physical" complexity of multicomponent systems (generally with components of heterogeneous nature), as well as the "functional" complexity of systems with many different functionalities and possible points of view. This is the manufacturing systems domain (of which TLSs constitute a particular element), as is pointed out in the following Table 1, where different points of view for manufacturing systems are highlighted, along with the corresponding design objectives, typical design solutions, and expertise domains. It may be noted that establishing manufacturing system design methodologies becomes a matter of linking together heterogeneous solutions proposed by different expertise domains, ranging from optimization models to heuristics, and from rule-based systems to component databases.

Traditional design theories are difficult to apply, due to the characteristics that the design process assumes when dealing with such functional and conceptual complexity. For example, it is possible to discuss a typical representation of design processes, based on the procedural paradigm, in order to notice its limited usefulness in the context of complex system design (see flow chart in Figure 5, [15]). With procedural models, establishing a design methodology is seen as equivalent to defining a more or less deterministic sequence of design steps, together with a set of procedures for carrying out each of them; given such methodology, designing means to follow this sequence in order to define system parameter values. The procedural approach gives great importance to procedures and procedure sequencing, while system parameters are just seen as the natural resulting outputs.

Now, as a first consideration in the discussion of traditional approaches compared to complex system design, it can be pointed out that the latter is a discrete decisional process, since the design parameters that are progressively assigned generally are of discrete nature and, therefore, their values influence the choice of the following design steps.

Table 1. "Functional" complexity of manufacturing systems.

Points of view	Design objectives	Typical design solutions	Expertise domain
Logistic network to be structured	Determine capacities of logistic elements, their topological connections and physical location	Mathematical models, heuristics	Operations Research
Logistic network to be managed	Choose and implement management strategies	Mathematical models, heuristics	Operations Research
A set of technological processes	Choose and design manufacturing processes, integrate them	Handbooks, heuristics	Manufacturing Engineering
A set of components to be assembled	Choose commercial components, design custom ones, determine interfaces	Handbooks, Databases and Catalogues	Manufacturing and Mechanical Engineering
A network of informative elements	Define software and hardware architecture of the system informative support	Handbooks, Software Engineering Methodologies	Information Science, Software Engineering, Electronic Engineering

As an example, in a transportation system the continuous value "number of objects moved per time unit" brings a discrete choice of component types (e.g., conveyors, robots, AGVs, etc.), and each component type will be characterized by design parameter values which are peculiar to each. Moreover, the choice of a component type will influence the choice, design, and implementation of management strategies for the same (in the example, none for conveyors, simple schedulers for robots, more complex schedulers for AGVs).

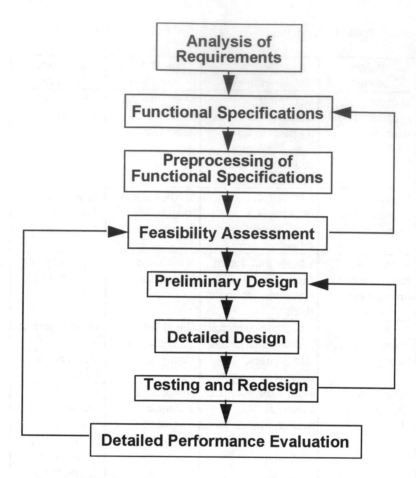

Figure 5. Design Process Flow Chart.

The discrete nature of the design process is also increased by the following:

- depending on the case, design parameters may be known in advance or not (thus becoming in the first case unknowns to be determined, and in the second case either input data or constraints to be satisfied);
- availability and reliability of input data may vary from one design to the other;
- the designer may have to parse the design of a set of systems in parallel; each system originates when, at a certain step, more than one option must be taken into consideration. This typically happens because cost/benefit tradeoffs often are uncertain at the moment of the choice.

For these reasons, manufacturing design systems may be depicted (Figure 6) as a maze with many entrances and many exits (or, in more scientific terms, as a decisional tree where it is possible to enter at different nodes). It is clear that traditional flow chart representations of the design process as in Figure 5 aren't able to adequately support the modeling of the design process, since it is rather pointless to specify strongly constrained design paths.

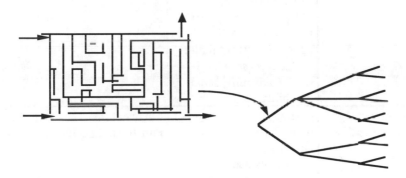

Figure 6. The discrete nature of the design process in manufacturing systems.

3.2   BREAKDOWN OF THE PROBLEM
INTO MAIN DESIGN PHASES

The TLS design problem, which has been overviewed in Section 2, is now to be challenged by keeping in mind the previous considerations. A first attempt in this direction may be to detail the phases shown in Figure 5, in order to relate TLS design to the more traditional models of design. In Table 2 each of these traditional design steps is associated with respect to the problem of designing a Tool Logistic System. The third column contains the design activities actually carried out by the system designer when the proposed design methodology is to be used. Each of these activities require the development of a design support software module implementing a rule, algorithm, or procedure. The contents of the modules shown in the last column will be discussed in Section 3.3. However, the way with which design activities are to be *managed* in the context of the proposed methodology departs from the deterministic sequencing of Figure 5, and will be described in Section 3.4.

Table 2. Breakdown of the TLS design process steps.

Standard Design Activities	Standard Design Activities in TLS design	Design Activities to be Performed (According to the Proposed Methodology)
Analysis of Requirements	Requirements definition: - cost of TLS - required efficiency (MC utilization rate, system reliability of FMS, operating costs of tooling) - effectiveness of TLS (tool availability, setup flexibility)	Carry out interview and edit Requirements List
Functional Specifications	Quantitative evaluation of requirements Definition of the constraints due to the actual scenario being addressed: - process plans - machine layout and component description - production data (mix/batch) - tool transport batch size - manned/unmanned operation	Carry out interview and edit Functional Specification List
Preprocessing of Functional Specifications	Grouping of tools and parts into families Translation of production data into production plans	Apply Group Technology like methods Based on layout and production data, generate a tentative workload schedule over a set of tool/part families
Feasibility Assessment	Preliminary drafting of the TLS graph, Capacity estimation of tool storages and transports	Evaluate feasibility of real-time tool sharing, Load parts on MCs through balancing of tool needs Estimate TLS storage and transport capacities

Preliminary Design	Selection of a preliminary TLS layout	Verify necessity of adopting tool buffers Evaluate feasibility of joining tool and part transport devices Verify utility of local tool transports Select corresponding standard layout
Detailed Design	Selection of TLS management strategies	Compute the optimal schedule of parts and tools, given the selected layout
Testing and Redesign	Verification of the conformity of obtained design with the initial specifications, and design modifications	Evaluate the level with which requirements are satisfied Modify the TLS design
Detailed Performance Evaluation	Final design validation through simulation	Carry out simulations of the designed TLS

In developing the methodology, it has been found convenient to aggregate these steps into three "macro" design phases. This first for the intuitive objective of grouping design activities oriented to the definition of system attributes of a similar kind. The second reason derives from the fact that designing a TLS requires to find a tradeoff among variables such as TLS transportation and storage capacity, system and tooling costs, and overall FMS performance. It is clear that this design is a complex, multiobjective problem. A practical approach in solving it may be to transform some objectives into constraints and to solve the corresponding set of single-objective subproblems. Subproblem solutions may be iterated until a satisfactory compromise is obtained. The aggregation of design activities in "macro" phases allows the designer to conveniently iterate among them during the design process (see Figure 7).

The proposed *Design Phases* are the following:

- TLS Structure Generation,
- TLS Management Optimization,
- TLS Efficiency Estimation/Evaluation.

The relationships between these design phases and the design steps which have been previously introduced are given in Table 3.

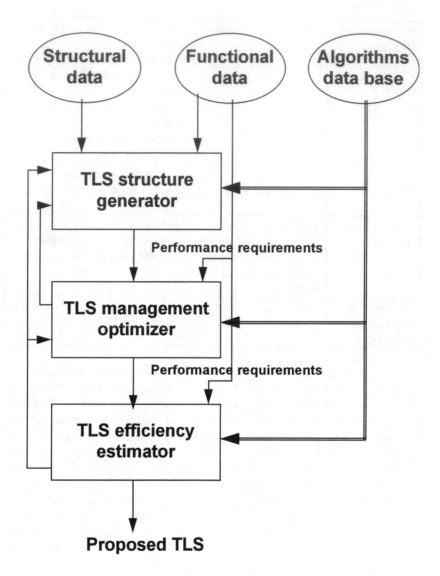

Figure 7. TLS-oriented design phases.

Table 3. Design phases with respect to design steps.

Design phases	Design steps
TLS structure generator	Analysis of requirements Functional specifications Pre-processing of functional specifications Feasibility assessment Layout selection by preliminary design
TLS management optimization	Detailed design for TLS functionality optimization
TLS efficiency estimation/evaluation	Testing and redesign Detailed performance evaluation

3.3 DESCRIPTION OF THE TLS DESIGN PHASES

3.3.1 TLS structure generation

The aim of the TLS structure generator is to lead the designer from the initial data available up to a preliminary design of the TLS architecture. This is performed according to the following steps:

Analysis of requirements: the TLS designer must settle the cost/ efficiency/effectiveness bounds which the TLS has to conform to.

Functional specification: the TLS designer has to discuss with the FMS end user the constraints concerning system layout, commercial TLS components, the production scenario, together with particular requirements concerning modes of production (e.g., whether the end user is willing to operate daily unmanned shifts). The interaction between designer and end user must point out peculiar specifications desired by the latter.

Functional specifications preprocessing: first, the TLS designer has to evaluate the convenience of referring to part and tool families for the whole TLS design, therefore reducing the number of variables (this activity is performed according to principles derived from Group Technology). Second, he has to translate functional specifications concerning production into a tentative production plan; this will allow finding a first attempt tool loading condition for the MCs.

Feasibility assessment: this phase allows estimating tool flow towards MCs, and tool storage level on the same, thereby estimating the capacity required by the tool storage and transport functions. This is done on the initial hypothesis of working with the most simple of the TLS layout configurations available (the one coded 1A), with no tool buffers and shop floor magazines, line-line configuration and disjoint part and tool transport. Within this step the following activities are developed:

- evaluate feasibility of real time tool sharing (i.e., the convenience of utilizing a tool on more MCs by moving it directly from one MC to another).
- load parts on MC through balancing of tool needs: workloads are assigned to MCs depending on production plans, they are then translated into a required tool amount. The latter is balanced over the MC set, thus obtaining a machine loading scenario which is balanced according to tool needs. This activity may be performed in different ways, whether production data has been previously defined as a repetitive production scenario or a batch one.
- dimensions of TLS storage and transport capacities: this is done by computing the number of utilization requests for each tool issued by MCs, depending on the assigned workload.

Preliminary design: in this step the structure derived at the preceding step is modified, and a preliminary system layout is selected through the following activities:

- verify necessity of adopting Tool Buffers (in case the number of Tools to be placed on some Tool Magazine should considerably exceed the capacity of commercially available Tool Magazines);
- evaluate feasibility of joining tool and part transport devices (by comparing transport frequencies of tools and parts, provided that utilization of dedicated devices is low);
- verify usefulness of local tool transport systems;
- select layout configurations (on the basis of the decisions taken in the preceding activities, one of the prototypical layout configurations may be identified);

These results correspond to the definition of a graph representing the topological links between MCs for tool storage and supply.

3.3.2 TLS Management Optimization

This phase corresponds to the optimization of the TLS functionality through definition of management strategies. Optimal scheduling of parts and tools is obtained through the choice of one among available algorithmic tools, depending on the production requirements and the TLS structure. The purpose of TLS Management Optimization is to provide the designer with a library of joint tool/part scheduling strategies in order to obtain maximum performance from the FMS. Indeed, a key point of the proposed TLS Design Methodology is that the TLS design must include the definition of a suitable management strategy.

The definition of the management strategy can be carried out only after the TLS structure has been completely generated. However, some functions of the TLS management optimization also play a role within other design phases, since :

- machine loading modules are used by TLS structure generation;
- scheduling strategies suitable for tool sharing can be used in order to spot bottleneck tools in a dynamic setting, thus influencing TLS structure generation;
- quality of obtained schedules can be used as an *estimate* of the TLS efficiency.
-

In the context of TLS design, the definition of a management strategy is complicated by a number of issues, such as:

- the need to coordinate part and tool flows;
- the variety of possible scheduling objectives (such as meeting due dates, maximizing machine utilization, etc.);
- the variety of constraints to be satisfied (limited availability of fixtures, pallets and tools; constraints on tool storage);
- the variety of TLS configurations.

A natural consequence is that there is no "best" strategy, and the task of the methodology is to assist the designer in progressively restricting the choice to the strategy most suitable to the TLS structure at hand.

It is known that the core problem of FMS management is machine loading, which consists of assigning operations and tools to machines in order to optimize a given objective function. As remarked in [22], the machine loading problem bears a great similarity to the bin-packing problem, formulated as the following:

a) given a set of items of given "size", and a set of bins, assign items to bins in order to distribute the load as evenly as possible (to be dealt with when balancing the load over MCs)

b) given a set of items of given "size", and bins of given capacity, assign items to bins in order to minimize the number of bins used (related to the batching problem, which must be solved when the FMS isn't able to support production of all part types concurrently; part type selection must be solved by batching together parts with similar tool sets, see [13]).

A multidimensional bin packing problem must actually be coped with, where the two dimensions are machine workload and tool magazine space. In machine loading one has to balance workloads under tool magazine constraints, while when estimating tool magazine sizes, one has to balance tool allocation over the machines under load balancing constraints.

In order to form a preliminary classification, approaches to FMS management which have been provided in the TLS design methodology may be grouped in three families as follows:

Joint tool/part scheduling: the two subproblems of tool and part scheduling are jointly solved. Although it is theoretically possible to build mathematical models to solve this problem, the only practically viable approach is to use priority rules.

One-way hierarchical approaches: in this class one of the two flows is subordinated to the other; two subclasses may therefore be distinguished, that is "tools-first" and "parts-first". In the former, tools are loaded on machines (e.g., in order to balance workloads) and then parts are scheduled, taking into account machine capabilities determined by tooling. The tools-first approach is usually associated to TLS with poor transportation performance: in such a case the FMS is loaded with a set of tools and selected part types are produced; then the system is shut down and tool setup is changed. The scheduling horizon is therefore divided into loading conditions at the end of which the whole set of active part types is changed. A flexible approach [38] may be appropriate, whereby loading condition variation is represented by the start or completion of a single part type. This, however, requires the adoption of a more powerful TLS, which can be exploited by a parts-first approach. In the parts-first approach, the part schedule is computed, then tool schedules are determined in order to obtain the least possible degradation of the part schedules.

Two-way hierarchical approaches: in this third class, one of the two flows is again subordinated to the other, but the "master" one may be revised in order to improve the "slave" one.

Although the management problems may be cast in the form of discrete optimization problems, their theoretical computational complexity, as well as their practical complexity in terms of implementation difficulties, discourages an optimization approach. Heuristic solutions have been preferred in the development of the methodology, being relatively easy to implement, understandable by nontechnical users, flexible, and adaptable to different problems. In addition, they allow interaction with users and are able to yield good solutions with reasonable computational effort. In the context of the methodology, heuristic solution approaches have therefore been chosen [4] among greedy procedures, myopic trajectory tracking methods, and local search.

TLS Management Optimization requires selecting, among the available management strategies, those compatible with the TLS Structure, by screening the suitable management strategies as a function of the TLS Structure attributes. Some strategies may be immediately ruled out, while the choice among the remaining ones is left to the designer, considering computational burden, flexibility, and implementation difficulty. When selecting a strategy the following discriminating features must be considered:

- is the TLS able to share tools among the machines?
- is it necessary to shut down the machines in order to change the tooling?
- is the production scenario repetitive or not?
- is the FMS based on a peculiar architecture (e.g., two-machine cells)?
- is the TLS Storage Architecture based on Tool Magazines only, or are Tool Buffers included?
- are due dates an important issue, or is FMS efficiency the most important scheduling objective?

3.3.3 TLS Efficiency Estimation/Evaluation

The objective of this last design phase is to provide, at first, a verification of the conformity of the obtained TLS with initial specifications. This is done by estimating the level within which requirements are satisfied and by modifying TLS design in accordance with the previous analysis. Finally, a detailed performance evaluation brings, through simulation, final design validation.

The purpose of such evaluations is to:

- check the feasibility of a design;
- rank different TLS alternatives (both in terms of structure and management strategy); it is worth noting that efficiency evaluation can be carried out on more than one TLS instance at once, allowing multiple criteria comparisons among competing design alternatives;
- improve the design by fine tuning quantitative attributes (e.g., tool magazine size or transportation capacity) or qualitative ones (e.g., the selected dispatching rule for scheduling);
- evaluate robustness of the design with respect to changing production scenarios and/or to unpredictable events;
- evaluate cost/efficiency tradeoffs.

To this purpose a number of possible efficiency metrics may be used: some of them must be evaluated by a thorough simulation of the TLS, thus requiring the use of a detailed simulation model. However, other metrics do not need simulation, since some of them are qualitative in nature, and others can be *estimated* from the solutions given by scheduling algorithms. It is therefore possible to distinguish between TLS efficiency *evaluation* and *estimation*: the former one is associated to traditional trial and error design methods based on simulation; the latter allows to "short circuit" this time-consuming design loop of "design, simulate, redesign". The ideal design process would be characterized by efficiency estimation for different TLSs, and only one simulation for the last one.

Efficiency metrics may be related to the performance during a simulation run for concerns:

Part performance: number of required parts, number of finished parts, throughput, flow time. For each part, time spent on MC table, process storage, handling.

MC performance: MC idle time (especially due to tool unavailability).

Transport performance: idle time, down time, shuttling (busy), loaded moving, unloaded moving, number of required missions, number of executed missions, mission time.

Tool storage devices: rate in, rate out, permanence time per tool, waiting time for services.

Tool type performance: for each tool type, time spent on spindle, time spent on tool magazine, time spent on handling devices, number of tools needed on MCs during simulation time.

Qualitative and economic performance attributes can also be considered. Qualitative attributes to be considered in TLS design include the following, and may be assessed during TLS structure generation.

Control complexity: this issue must be considered both for TLS Structure Generation (some TLS transportation devices require complex control software) and TLS Management Optimization (Tool Sharing policies are more complex to implement than Part Type Selection and Batching).

Expandability: i.e., how easy is it to expand the TLS when a Machining Center is added? This feature must be considered at the TLS Structure Generation Level (e.g., an AGV based system is easier to expand than a Gantry Robot based one).

Reliability: here we don't intend a quantitative reliability measure, but rather the TLS' ability to provide substitute tools for worn-out or broken ones.

The first two attributes do not require simulation, while the third one does if its quantification is required.

Economic attributes are basically related to cost issues, where TLS cost depends on components costs (i.e., tool magazines, transportation devices) and tooling costs (i.e., related to the number of physical tool copies required). Components cost is assessed during TLS Structure Generation; tooling cost is to be carefully considered when management policies based on tool sharing are used, since in this case trade-offs between tooling cost and operational performance indicators are to be evaluated. For this reason tooling costs, which may be assessed during TLS structure generation, are to be studied during the final simulation phase.

Apart from those previously listed, other operational attributes may be estimated from scheduling results, mainly makespan, mean weighted tardiness, maximum lateness, mean weighted flow time. It is worth noting that makespan is related to machine utilization, the flow time to work in process, tardiness and lateness to customer service.

More specific performance attributes can also be considered (more or less correlated one to another):

- the ability of the TLS to cope with different part type requirements simultaneously;
- the setup flexibility, i.e., the ability to change tool setup easily while maintaining good machine utilization;
- the number of batches required to carry out a given set of production requirements;
- the ability of balancing workloads.

Of course, the results emerging from scheduling results are more inaccurate than the ones emerging from simulation, nevertheless, useful information can be obtained in early stages of design, since when using simple strategies based on dispatching rules, information is obtained quickly. Assessing such attributes early without the need of extensive simulation runs allows reducing the time needed to design the TLS, or, by the same token, it enables the designer to broaden the number of TLS alternatives considered, ultimately leading to improved systems.

3.4 A BLACKBOARD BASED APPROACH TO MANAGEMENT OF THE DESIGN PROCESS

In the previous paragraphs TLS design data and functions have been introduced and overviewed; it is now necessary to define the principles of the design process management logic (DPML). The objective of such logic is to aid the designer in tackling specific TLS design problems, by starting from specific data and constraints and by applying available modules, algorithms, and rules to obtain the most appropriate design path, thus leading from initial data down to design completion. The DPML is to aid the designer in performing the three design phases:

- select a limited set of TLS configurations among the prototype layouts, progressively identifying new characteristics, leading to the unambiguous identification of a layout type;
- select from the available management techniques and for each of the configurations chosen, the best procedures for tool scheduling, monitoring, and control. The DPML has to aid the designer in the use of optimization algorithms and in comparing their performance, and to use the feedback given by him in order to direct the selection. In this way an iterative search for the best management techniques may be performed.
- assess the quality of the TLS structure and the management strategy chosen, with respect to the values assumed by a given set of indicators.

The DPML must therefore support the iterative definition and selection of TLS structures and management techniques, carried out by working on the necessary tradeoffs to be found between cost and performance.

In all three design phases it has been stressed that design involves a "progressive selection of attributes" at first concerned with the physical structure and topology, secondly to the decisional architecture, and in the end to the tradeoff analysis between cost and performance. The need for such "progressive selection" leads to the necessity of characterizing the DPML so that this specific kind of design process may be performed. The operating requirements cast on the DPML may be summarized as follows:

- it must constitute an "integration platform" for decisional procedures and tools of heterogeneous nature (mathematical models, rules derived from experience, knowledge of commercially available components, etc.);
- it has to be centered on the task of progressively selecting the appropriate methods for determining unknown parameters of the system (regarding both its physical and decisional architecture). In software engineering terms it may be thought of with an object-oriented approach, since the application of design procedures is to be controlled by the values assumed by object parameters (as, in object-oriented languages, method application is controlled by events related to the objects). Object orientation is here opposed to procedural orientation, since it has been discussed that the design process can't, for complex systems such as TLSs, be managed by rigid flow chart-like mechanisms;
- it has to allow the designer to carry on a design in a context where the variety and reliability of input data and constraints may differ from case to case.

In order to represent the design process resulting from such a scenario, it is possible to adopt the "blackboard model" used in the field of Artificial Intelligence [31]. Such model has been used for design support systems [16,

29], although never in the field of manufacturing system design. Blackboard systems are constituted by cooperating entities with decision-making tasks (termed knowledge sources), who communicate among themselves by writing and reading on a common data base (the blackboard). In order to apply the blackboard model to the system design domain, it is possible to think of a meeting being carried out by a group of designers, each of which has his own expertise and set of skills. A blackboard placed in front of the designers lists the attributes of the system being designed, and the agents may communicate only by writing and reading on the blackboard. Each designer's skill will enable him to intervene in the meeting and write on the blackboard some attribute value, provided that the data he needs for doing it has been previously written down by someone else.

Designer intervention may be controlled by behavioral models embedded in each agent, as well as by an external controller entity, the former solution being closer to the standards of Distributed Artificial Intelligence. The latter has, however, been adopted since the controller entity coincides in fact with the Design Process Management Logic, while the role of each "designer" is covered by one among the various decisional modules in charge of performing individual design steps.

The skills of the modules/designers may be summarized with a "blackboard access table" (see Table 4), in which the system attributes (the blackboard's "slots") are linked to the modules' characteristics.

Table 4. Blackboard access table.

	Modules	Module 1	......	Module n
Blackboard slots				
Attribute 1		r | w | p		
.....				
Attribute m				

In the access table one can see, for each module, which of the attributes:
- serve as input data ("r"), and are read by the module,
- are written as an output of the module ("w"),
- may influence (through their value) the possibility of activating the module. Certain modules may be activated only if some attribute values are of a specific kind ("r").

The DPML may be modeled by a finite state machine, whose state is defined by the attribute values, and whose transitions are given by module activation. Together with the attribute and module sets, it is necessary to define a controller which, at each design step, will read the current state and suggest to the designer which modules should be activated in order to reach a desired state (in particular the one corresponding to the completion of the TLS design). The mechanism may be clarified by Figure 8, in which it is possible to understand the roles covered by the DPML controller and by the designer, together with their mutual interactions.

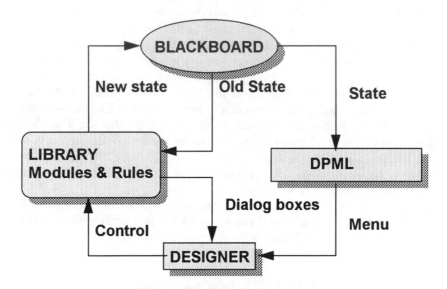

Figure 8. Interaction between the designer
and the Design Process Control Logic.

It is also possible to describe the DPML's operation in mathematical terms. The following mathematical formula may be used for:

• **the system being designed**
- let $A = \{a_1...a_n\}$ be the *attributes* of the system;
- let $V_j = \{\emptyset, v_{j1} ...v_{j\ nj}\}$ be the set containing the types of values that attribute a_j may have. By value type we mean the different values that may influence the designer in choosing alternative design modules. Attribute value $\emptyset$ means the attribute is still undefined;
- the space state is given by $V = V_1 x .... x V_n$ where $N = \Pi_j\ n_j$ states — or *configurations* — exist. Each configuration C is identified by an n-element vector, whose j-th element $c_j \in V_j$ contains the value of the j-th attribute,

- a configuration is *final* if all of its attributes are defined: $c_j \neq \emptyset \; \forall \; j = 1...n$.
- it is possible to define a set of final configurations $CF = \{C^* \mid c^*_j \neq \emptyset \; \forall \; j = 1...n\}$

• **the design procedures**
- let $P = \{p_1...p_m\}$ be the set of design *procedures*, able to define the system's attribute values,
- a procedure p_i is defined by $p_i = \{AF_i(C), IN_i, OUT_i\}$ where:
* $AF_i(C): AS \rightarrow \{0, 1\}$ is a Boolean function whose value is "1" if, starting from configuration C, it is possible to activate procedure p_i
* IN_i is an n-element Boolean vector, whose j-th element is "1" if procedure p_i needs to read attribute a_j
* OUT_i is an n-element Boolean vector, whose j-th element is "1" if procedure p_i determines attribute a_j
- Procedures must follow the following two rules:
* if a procedure reads an attribute, function AFi must require that it is written beforehand,
* a procedure may write an attribute only if it is undetermined, or if it has been read (in this case we have a *refinement procedure*). This is necessary in order to avoid that some attribute may be arbitrarily overwritten during the design process. If it is necessary to iteratively refine an attribute value, this may be done either with a refinement procedure, or by having the designer explicitly erase the attribute.
- An attribute j* is *undefined* if no procedure able to write it exists.

• **the design process**
- a *design* is a sequence of procedures $p_{s(1)}...p_{s(v)}...p_{s(q)}$ which starts from an initial configuration $C_0 \in V$ and leads to a final configuration $C_q \in CF$. At each step v, given C_v, it is necessary to choose a procedure able to lead to the next configuration:
$s(v+1)$ s.t. $AF_{s(v+1)}(Cv) = 1$,
- A configuration C is *dead* if it isn't final, but it doesn't allow progress in the design, if not by erasing some attribute value.

With this formalism, the definition of a design methodology is equivalent to defining:

- the attribute set A, the space state V, and the set of initial configurations from which a design task is likely to start,
- the procedure set P,
- a *design control strategy*, that is a mechanism for leading from any initial configuration to a final configuration. Such a strategy may be implemented in various forms, ranging from simple rules down to sophisticated algorithms for shortest path finding. A simple control strategy may be given by the following rules:
* a module is allowed to start only if all the TLS attributes to be read are filled in,
* at each step the DPML provides the end user with a "menu", listing the modules which, at the moment, the end user is allowed to activate,
* the designer's role is to choose, among the proposed modules, the one to be activated. This choice specifies the "control" of the end user on the TLS design decisional logic,
* the operation of a module after the end user's choice causes the transition to a new configuration of the TLS design process.

The difference of such logic with respect to the one implied by procedural design models is self-evident.

Given the methodology as defined above, the development of a specific design is equivalent to the following:
- identify the initial configuration,
- determine a sequence of procedures to activate by using the design control strategy by using the rules given by the defined design control strategy.

3.5 Structuring of TLS design knowledge
with the proposed DPML architecture

It is now possible to put together the technical contents of TLS design, as emerging from paragraph 3.3 with the DPML architecture of paragraph 3.4, in order to yield the global architecture of the methodology, as has been implemented in a computerized Decision Support System. The first step in this direction is given by the identification system attributes (or "blackboard slots") which are relevant both in terms of characterizing the system as such, and for directing the design process' trajectory. The results of such analysis may be seen in Table 5, along with a brief explanation for each.

Table 5. Meaning of TLS attributes.

TLS attribute	Meaning
Unmanned	expresses whether plant operation in unmanned shifts is to be foreseen
Tool requirements	data structure expressing global tool requirements
Mix	expresses whether FMS/FTL operates a repetitive production
Sharing	data structure expressing for which tools real-time tool sharing may be performed
MC Loading	data structure containing MC workloads and schedules
TM Capacity	data structure containing the size of Tool Magazines
THS Capacity	data structure containing tool flow intensity on the TLS graph
Batch	expresses whether the TLS manages tools transport in batches
Tool Batch Dimension	expresses size of the tool transport batch
Buffer	expresses whether Tool Buffers are to be provided
SFM	expresses whether a Shop Floor Magazine is to be provided
Joint	expresses whether the transport of parts and tools is to be performed by a common device
Tool Cell	expresses whether local tool transport systems are to be provided
Part Cell	expresses whether local part transport systems are to be provided
Layout	code identifying the selected prototypical configuration
Management Strategy	code identifying the selected management strategy
Quantitative perf. measures	data structure containing the set of operative performance measures
Qualitative perf. measures	data structure containing the set of qualitative performance measures
TLS Cost	data structure containing the cost of the selected TLS architecture
Cost-related perf. measures	data structure containing the set of cost-related performance measures

After having defined the blackboard slots (i.e., the TLS attributes), the existing TLS design knowledge overviewed in the preceding paragraphs may be inserted in the definition of design modules able to operate on attributes. The design modules identified are defined in Table 6.

Table 6. Specification of design modules.

Module name	Module description
M1	**Prerequisites:** TLS configuration without tool buffers, no tool sharing **Input:** technological and production data **Output:** machine loading for each MC. **Objectives:** to determine a first-attempt machine loading condition. Used during Structure Generation **Specification:** parts and tools are loaded so as to minimize maximum tool magazine size, while keeping machine workload as balanced as possible. Implemented with heuristics
M2	**Prerequisites:** TLS without tool buffers, no tool sharing **Input:** technological and production data, tool magazine size **Output:** number of required batches; for each batch: machine loading and part types included. **Objectives:** can be used both for Structure Generation and for Management Optimization. It implements a hierarchical one-way approach of tools-first type, and must be coupled with a part scheduler. Can be used both for part type selection and batching. Common objectives are workload balancing or maximization of the value of selected part types **Specification:** the module is implemented as a three step heuristic procedure
M3	**Prerequisites:** TLS configuration without tool buffers; the TLS must support tool sharing **Input:** technological and production data, tool magazine size for each MC, tool availability for each type. **Output:** for each MC, parts and tool assigned. **Objectives:** assignment of parts and tools to machines in order to minimize tool traffic. Used in Structure Generation and Management Optimization **Specification:** optimization algorithm or heuristics; the decision variables are part and tool assignment to machines; the constraints considered are tool magazine sizes, tool availability and workload balancing.

M4	**Prerequisites:** No TLS architecture is assumed. It is the most general management module, both in terms of TLS Structure and scheduling objectives.
	Input: technological and production data, TLS structure (optional)
	Output: part and/or tool schedule.
	Objectives: to provide well-known priority-based approaches to FMS scheduling. All TLS structures are covered. For cases covered by other modules it enables verifying whether more sophisticated approaches can yield advantages. Used in Structure Generation and Management Optimization
	Specification: a rough discrete event simulation of dispatching rules
M5	**Prerequisites:** a FMS based on two-machine cells with common tool buffer
	Input: Production and Technological data, tool buffer size for each cell, tool availability.
	Output: part and tool schedule for each cell and MC.
	Objectives: minimization of the makespan by balancing workloads among cells and by minimizing conflicts on common tools within each cell. Used in Management Optimization
	Specification: a two-step procedure based on heuristics
M6	**Prerequisites:** being based on a continuous flow approximation, it is *not* suited to very small batches.
	Input: technological and production data,
	Output: cumulative production trajectories, specified as parts to be produced during each time bucket.
	Objectives: to create a cumulative production trajectory over a discrete time horizon. Production trajectories must then be tracked; real time tool sharing and buffering can be exploited. Used in management optimization
	Specification: mathematical programming model, solved with Linear Programming algorithms.
G1	**Prerequisites:** none
	Input: technological and production data
	Output: tool sharing, tool requirements
	Objectives: assessing which tools are suitable (because of cost and frequency of use) for real-time sharing, and deciding number of physical tools to use. Used during Structure Generation
	Specification: three step heuristic procedure

G2	**Prerequisites:** none **Input:** technological and production data **Output:** part/tool groups **Objectives:** to assess opportunity of working on part families/tool groups, and to find them. Used for Structure Generation **Specification:** three step heuristic procedure
C1	**Prerequisites:** TLS configuration without tool buffers, and mix-oriented production **Input:** loading condition, kind of operation **Output:** Tool magazine size **Objectives:** used during Structure Generation **Specification:** simple computation
C2	**Prerequisites:** TLS configuration without tool buffers, and batch-oriented production **Input:** loading condition, kind of operation **Output:** tool magazine size **Objectives:** used during Structure Generation **Specification** simple computation
C3	**Prerequisites:** TLS configuration without tool buffers, no tool sharing, constant mix production **Input:** loading condition **Output:** required TLS transport capacity **Objectives:** used during Structure Generation **Specification:** rough simulation
C4	**Prerequisites:** TLS configuration without tool buffers, and tool sharing allowed, constant mix production **Input:** loading condition **Output:** required TLS transport capacity **Objectives:** used during Structure Generation **Specification:** rough simulation
C5	**Prerequisites:** TLS configuration without tool buffers, batch production **Input:** loading condition **Output:** required TLS transport capacity **Objectives:** used during Structure Generation **Specification:** rough simulation
R1	**Input:** Tool magazine capacity, TLS transport capacity, type of operation, **Output:** opportunity of using buffers, shop floor magazines, joint transportation, tool cell transport, layout code **Objectives:** used during Structure Generation **Specification:** rule-based procedure

R2	**Prerequisites:** Completion of Structure Generation **Input:** TLS attributes **Output:** feasible management techniques **Objectives:** used during Management Optimization **Specification:** rule-based procedure
E1	**Prerequisites:** Completion of Structure Generation **Input:** TLS attributes **Output:** cost of TLS, qualitative performance measures **Objectives**: used during Efficiency Estimation **Specification**: simple computations
E2	**Prerequisites:** Completion of Structure Generation **Input**. Cost of TLS, operative measures **Output**: cost-related performance measures and tradeoffs **Objectives** used during Efficiency Estimation **Specification**: simple computations
E3	**Prerequisites:** Completion of Structure Generation and Management Optimization **Input:** TLS attributes **Output**: Operative performance measures **Objectives**: used during Efficiency Estimation **Specification**: detailed simulation model

Finally, the relationships between these design modules and TLS attributes may be synthesized by Table 7, through the definition of modules' read/write accesses to the blackboard.

It is assumed that those TLS attributes which can only be read by modules have to be filled by the end user, or read by the data base containing process plans and global production data. When all of these attributes have been defined, the TLS design decisional logic can start its activity, based on the rules expressed in paragraph 3.4.

The TLS design methodology is currently being implemented as a computerized Decision Support System, developed following the blackboard architecture, and integrating different software modules implementing each of the previous design modules.

The user interface of the system is shown in Figure 9, where it is possible to see both the blackboard which gives the status of the design being performed and the TLS' attributes, and the module menu (upper left) showing, at each design step the applicable modules.

Table 7. TLS attribute blackboard access table.

Blackboard	M1	M2	M3	M4	M5	M6	G1	G2	C1	C2	C3	C4	C5	R1	R2	E1	E2	E3
Unmanned									r	r				r-w	r			r
Tool Req.'s							w								r			r
Mix	p	p-w							p	p	p	p	p	r-w	r			r
Sharing	p	p	p					w			p	p	r	r-w	r			r
MC loading	w	w	w	w	w	w	w		r		r	r	r	r				r
TM capacity		r	r	r	r				w	r	w	w		r	r	r		r
THS capacity										w	w	w	w	r	r	r		r
Batch														w	r			r
Tool batch dim.														w				r
Buffer	p	p	p	r	p									w	r	r		r
SFM				r										w	r	r		r
Joint				r										w	r	r		r
Tool cell	p	p	p	r	p									w	r			r
Part cell				r										w	r			r
Layout				r										w	r			r
Mgmt strategy															w			r
Quant. perf. mes.	w	w	w	w	w	w											r	w
Qual. perf. mes.																w	r	
TLS Cost																w	r	
Cost-rel. perf. mes.																	w	

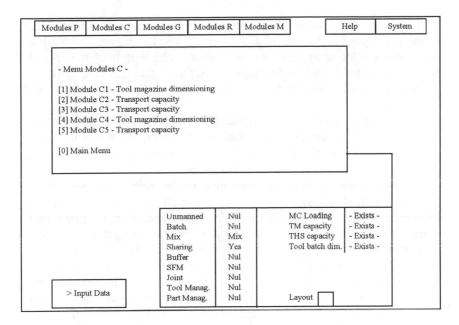

Figure 9. The design support system's user interface.

4. CONCLUSIONS AND DIRECTIONS FOR FURTHER RESEARCH

In this paper a methodology for designing a Tool Logistic System has been presented.

The methodology, conceived for providing a designing tool to FMS experts, has been structured in a fairly interactive way, on the basis of author's skepticism about fully automated design methodologies. This means that the methodology has to be seen as formed by three different subjects: a basket of automated procedures, a guide for suggesting best procedures to be used at each stage of the design process, and human designer.

The latter remains the real engine of the system, in the sense that every single decision is left to human decision. Authors believe that this must be a milestone of the methodology, therefore it is not in the scope of further research to put any emphasis on the automation of the decisional kernel.

On the contrary, it is understood that the knowledge base has to be continuously improved. Under this perspective the system has been designed in a modular way, in order to allow a continuous addition of pieces of knowledge (see for instance module R1-R4).

This applies both to knowledge embedded in the specific procedures and to knowledge included in the data base which should support the whole

TLS design methodology (e.g., commercially available components, existing design, etc.). In particular, we consider this latter a point which has to be given a lot of attention in the future. The design of the logical structure of a technological data base, its implementation, and updating is actually a crucial point on the way of defining a truly effective design methodology.

Besides the intrinsic value of disposing of a rich data base, we hope that its definition will carry over some side effects on the definition of single procedures, with particular reference to the problem of enriching the information contents of specific blackboard slots.

The blackboard approach has appealed to the authors because of its similarity with the design process which should be dreamt of in every modern enterprise: a group of designers brainstorming around the table and putting their different skills (the procedures in the methodology) at the service of the project. In this perspective the blackboard becomes the "perfect project manager" which at every stage of the design process has the perfect information of the state of the project and of capabilities of his staff. Nevertheless, in reality just like in our methodology, such kind of situations become unmanageable when exceeding reasonable complexity and dimension. Preliminary tests have, in some way, confirmed this suspects. Therefore authors believe that a further important direction of research should be the one focused on the objective of structuring the blackboard in such a way that a substantial increase in the number of parameters to be defined and design modules won't affect methodology consistency and effectiveness.

As a last consideration, it has to be noted that the problem of drafting the actual floor plan is at the present state is yet to be developed. Leaving aside the implementation issues related to interfacing to CAD systems, which are out of the scope of the research, a truly interesting topic is represented by the definition of the logic with which the designer may avail himself in order to move from the layout code to the final instantiation of the TLS with actual components.

ACKNOWLEDGMENTS

The results discussed in the paper are based upon research work supported by the European Community under Research Contract BRITE-EURAM n. 4113.

REFERENCES

1. **Amoako-Gyampah K.**, **Meredith J.R.**, **Raturi A.** (1992), "A comparison of tool management strategies and part selection rules for a flexible manufacturing system", International Journal of Production Research, vol.30 n.4
2. **Arbib C.**, **Lucertini M.**, **Nicolò F.**, (1990) "Workload balance and part transfer minimisation in FMS", International Journal of Machine Tools Manufacturing, pp.5-25
3. **Berrada M.**, **Stecke K.E.**, (1986), "A branch and bound approach for machine load balancing in flexible manufacturing systems", Management science, vol.32, pp.1316-1335
4. **Brandimarte P.** (1992), "Neighbourhood Search-Based Optimization Algorithms for Production Scheduling: a Survey", Computer-Integrated Manufacturing Systems, Vol. 5, pp. 167-176.
5. **Calderini M.**, **Cantamessa M.** (1992), "Analysis of process plans for tool magazines' sizing in flexible manufacturing systems" unpublished BRITE-EURAM 4113-Internal Report
6. **Calderini M.**, **Cantamessa M.** (1993), "A representation framework for the analysis and design of tool logistic systems in flexible manufacturing", Proceedings of the AMST'93 International Conference, Udine (Italy), April 26-29th 1993
7. **Cantamessa M.**, **Lombardi F.** (1993), "Tool flow planning in a flexible manufacturing system", Computer-Integrated Manufacturing Systems, vol.6 n.2
8. **Carrie A.**, **Bititci U.S.** (1990), "Tools for integrated manufacturing", Proceedings of the 28th International MATADOR Conference, Manchester, UK, April 18-19th 1990
9. **Carrie A.**, **Pereira C.T.S.**, (1986), "Work scheduling in FMS under tool availability constraints", International Journal of Production Research, vol.24 n.6
10. **Chang Y.-L.**, **Matsuo H.**, **Sullivan R.S.**, (1989), "A bottleneck-based beam search for job scheduling in a flexible manufacturing system", International Journal of Production Research, vol.27, pp.1949-1961
11. **Chen-Hua Chung** (1991), "Planning tool requirements for flexible manufacturing systems", Journal of Manufacturing systems, vol.10 n.6
12. **Chen I.J.**, **Chung C.-H.** (1991), "Effects of loading and routing decisions on performance of flexible manufacturing systems", International Journal of Production Research, vol. 29, pp. 2909-2225
13. **Co H.C.**, **Biermann J.S.**, **Chen S.K.**, (1990), "A methodical approach to the FMS batching, loading and tool configuration problem", International Journal of Production Research, vol. 29 n.12

14. **De Vecchi L.**, **Parola G.**, **Semeraro Q.**, **Tolio T.** (1992), "TWM1: a system for integrated tool and workpiece management in FMS - loading and part selection", 24th CIRP International Seminar, Copenhagen, Denmark, June 11-12th, 1992

15. **Ertas A.**, **Jones J.C.**, "The engineering design process", John Wiley & Sons, 1993, p.3

16. **Garret J.H.**, **Lu S. C-Y**, **Thompson J.B.**, **Herman A.E.** (1990), "Applications of AI techniques to engineering design", 5th International Conference on Applications of Artificial Intelligence in Engineering, Boston, MA, July 1990, Springer Verlag

17. **Gray A.**, **Seidmann A.**, **Stecke K.** (1988), "Tool management in automated manufacturing: operational issues and decision problems", Working paper, University of Rochester.

18. **Gunasingh**, **Lashkari**, (1993), "Machine grouping problem in cellular manufacturing systems - an integer programming approach", International Journal of Production Research, vol.27 n.9

19. **Han M.-H.**, **Yoon K.Na**, **Gary L.Hogg** (1989), "Real-time tool control and job dispatching in flexible manufacturing systems", International Journal of Production Research, vol.27 n.8

20. **Ippolito R.**, **Levi R.**, **Lombardi F.**, **Tornincasa S.**, **Villa A.**, (1989) "High speed turning: technological and economical involved aspects", in Proceedings of the CNR/CIRP seminar on High Speed Cutting, Turin (Italy), June 1989

21. **Kasinlingam**, **Lashkari** (1993), "Cell formation in the presence of alternate process plans in FMS", Production Planning and Control. vol.2 n.2

22. **Kim Y.-D.**, **Yano C.A.** (1993), `Heuristic Approaches for Loading Problems in Flexible Manufacturing Systems", IIE Transactions, vol. 25, pp. 26-39.

23. **Kouvelis P.**, **Chiang W.C.**, **Kiran A.S.** (1992), "A survey of layout issues in flexible manufacturing systems", in OMEGA, International Journal of Management Science, vol.20 n.3

24. **Kouvelis P.**, **Chiang W.C.** (1992), "Design and planning problems in flexible manufacturing systems: a critical review", Journal of Intelligent Manufacturing, N.3

25. **La Commare U.**, **Noto La Diega S.**, **Perrone G.** (1992), "FMS management considering tool and part flow coordination", 24th CIRP International Seminar, Copenhagen , Denmark, June 11-12th, 1992

26. **Lambiase A.**, **Riemma S.**, **Santillo L.C.** (1993), "The optimization of tooling management in a flexible manufacturing system", Proceedings of the AMST'93 International Conference, Udine (Italy), April 26-29th 1993

27. **Leung L.C., Maheshwari S.K., Miller W.A.** (1993), "Concurrent part assignment and tool allocation in FMS with material handling considerations", International Journal of Production Research, vol.31 n.1

28. **Mandelli S.p.A**: (1991), private communication

29. **Medland A.J.**, (1992), "The computer-based design process", Chapman and Hall, London.

30. **Montazeri M., Van Vassenhove L.N.** (1990), "Analysis of scheduling rules for an FMS", International Journal of Production Research, vol.28 n.3

31. **Nii H.P.**, (1989), "Blackboard Systems", in Handbook of Artificial Intelligence, vol. IV, Chapter XVI, Barr A., Cohen P.R., Feigenbaum E.A. (Eds.), Addison Wesley,

32. **Nilsson K.** (1987), "Adaptive Tool System", Proceedings 6th International Conference on Flexible Manufacturing Systems, Turin, Italy, 4-6 nov. 1987

33. **Ram B., Sarin S., Chen C.S.** (1990), "A model and a solution approach for the machine loading and tool allocation problem in an FMS", International Journal of Production Research, vol.28, pp.637-645

34. **Sarin S., Chen C.S.** (1987), "The machine loading and tool allocation problem in a flexible manufacturing system", International Journal of Production Research, vol.25 n.7

35. **Shanker K., Srinivasulu A.** (1989), "Some solution methodologies for loading problems in a flexible manufacturing system", International Journal of Production Research, vol.27 n.6

36. **Srivastava B., Chen W.-H.** (1993) ``Part type selection problem in flexible manufacturing systems: tabu search algorithms", Annals of Operations Research, Vol. 41, pp. 279-297.

37. **Stecke K.** (1983), "Formulation and solution of nonlinear integer production planning problems for flexible manufacturing systems", Management Science, Vol.29, N.3

38. **Stecke K.E., Kim I.** (1991), "A flexible approach to part type selection in flexible flow systems using part mix ratios", International Jou. of Production Research, Vol. 29, pp. 53-75.

39. **Tang C.S., Denardo E.V.**, (1988a), "Models arising from a flexible manufacturing machine, part I: minimization of the number of tool switches", Operations Research, vol.36, pp.767-777

40. **Tang C.S., Denardo E.V.**, (1988b), "Models arising from a flexible manufacturing machine, part II: minimization of switching instants", Operations Research, vol.36, pp.778-784

41. **Tetzlaff U.** (1990), "Optimal design of Flexible Manufacturing Systems", Physica-Verlag, Heidelberg.

42. **Widmer M.** (1991), "Job shop scheduling with tooling constraints: a tabu search approach", Journal of the Operational Research Society, vol.42, pp.75-82

43. **Wilson J.M.** (1989), "An alternative formulation of the operation-allocation problem in flexible manufacturing systems", International Journal of Production Research, vol.27 n.8

44. **Zavanella**, **Bugini** (1992), "Planning tool requirements for flexible manufacturing: an analytical approach", International Journal of Production Research, vol.30 n.6

Chapter 8

A PROJECT FOR A LOGISTIC INTEGRATED NETWORK FOR DECISIONS ENHANCEMENT (LINDEN)

Walter Ukovich
Department of Electrotecnics, Electronics and Informatics
University of Trieste, Italy

Marco Crasnich and **Franco Zanetti**
IBM Semea S.p.A. - Vimercate (MI), Italy

1. INTRODUCTION

The role of logistics in production systems is undergoing an accelerated evolutionary process that leads it to cover an increasingly wider area. Adoption of innovative production philosophies, enhancement of production technologies, market globalization, tighter competition challenges, and rising attention to quality issues are certainly among the factors driving such an evolution.

As a consequence, decisions that logistic managers must face increase in number and complexity on one side, and in relevance of their impact on the other side. This produces in turn a growing need for tools capable of supporting these decisions.

This paper is devoted to outline an R&D project to design a tool aimed at providing an effective answer to such requirements. It is a Logistic Integrated Network for Decisions Enhancement (LINDEN). LINDEN is intended to be an **integrated network** providing tools for supporting the decisions of the logistic manager — and, more in general, all the decisions involving logistic issues — in each point of the **logistic network**, for all decision levels: strategic, tactical and operational.

The scope of this paper is twofold. First, it provides an opportunity to promote partnership in the LINDEN project from companies, institutions, and research groups that may have interest in supporting its activities and in having part in them. Second, it proposes a base for discussion among prospective partners on the contents of the LINDEN project and on the operations devoted to implement it. In this sense, this paper aims at stimulating comments, criticisms, contributions directed to improving its structure and contents, and to involving interested partners.

Accordingly, this paper is divided in two parts. The first part encompasses Sections 1 to 3. Here the structure of LINDEN is first sketched and its most relevant proposed features are pointed out. The characteristics of prospective partners in the project are outlined, and some motivations for them are provided. Finally, a rough estimate of the necessary resources is proposed. The second part consists of Section 4. It describes the project to design LINDEN more in detail and provides operational insights on how it is intended to be carried out.

Decision Making in Logistics

In this section we briefly summarize some of the most relevant characteristics of general decision making processes in logistics, which constitute the conceptual framework into which LINDEN has been conceived.

As a general definition of logistics, we take (cf. [10]) *the complex of all activities devoted to modify time and space attributes of any resource is managed within a production system*, such as materials, goods, products, personnel, labor, information, etc. Although this definition is slightly more general than some of the many provided in the literature (see for instance, among the others, Daganzo [1], Daskin [2], Graves et al.[4], Johnson and Wood [5], Hall [6], Hameri and Eloranta [7], Sims [9]), it seems to be quite adequate in order to help to discriminate between what we consider a logistic process and what we do not.

So in abstract terms, a production system (in its most general formulation: *any system* — organization, company, firm, agency, etc. — *devoted to providing products, or services, to customers* may be viewed as a complex of transformation processes, which in turn are fed by logistic processes with the resources they need to operate.

Such a simple model suffices to infer the most relevant characteristics of logistic processes:

- logistics involve processes that are *secondary* with respect to transformation processes, in the sense that they do not add value to the resources involved; in this sense "they add nothing but cost", as says Sims [9];
- however, the primary (i.e., transformation) production processes would die by starvation without logistics;
- moreover, logistics impacts on the *quality* of products and production processes, since time and space attrubutes are related to service levels;
- logistics *pervades* the production system: it fills the gaps between the transformation steps, and provides the means that make them work;

- as a consequence, logistics always deals with *multiactor* situations;
- hence a *multiattribute* attitude (in the sense of Keeney and Raiffa [8]) is intrinsic to logistics;
- summing up, *integration* is the most appropriate approach to decision processes in logistics.

The latter conclusion complies with the opinion of most authors in the field, possibly with different formulations, such as "the system approach" to logistics (see among the others Johnson and Wood [5], and Sims [9]). Therefore, in general terms, LINDEN must be a tool both to take decisions in an integrated way and to enhance integrated decisions.

Integration is achieved by exchanging information between the decision makers. We always distinguish two types of actors in logistic processes: *customers*, who require a logistic service, and *suppliers*, who provide the required logistic service. Depending on the specific situation, suppliers and customers of logistic services may belong to different companies, or different departments of the same company, or they may even coexist in the same structure.

What is fundamental is that decision makers of either type concur to carry out the logistic processes, and they have in general different (and possibly contrasting) criteria and objectives. Nevertheless, they agree to cooperate as *partners* in what we call a *contract* in order to carry out the logistic process. Therefore, in terms of the Total System Intervention classification (see Flood and Jackson [3]), logistic fits into the realm of the *complex* and *cooperative* situations.

Therefore decision processes in logistics may be enhanced by:

1. the effectiveness of the decision procedures;
2. the quality of the available information.

According to the above discussion, these requirements should be pursued by providing 1) the possibility of integrating in a flexible way several multicriteria models, and 2) a convenient support to exchange data. These constitute the basic features of LINDEN.

2. OUTLINE OF LINDEN

As written above, LINDEN is a network of decision aid tools devised to support decision makers in logistics. As a network, it is physically distinct from the *logistic network*, which is where resources actually reside and move. In this section we sketch the features of the LINDEN network.

2.1. Local elements

At the local level, the basic element of LINDEN is a *logistic workstation*, possibly operated by the decision maker, containing different modules to face his specific needs, and providing access to the relevant information for the specific decisions.

Essential features of the logistic workstation are:

- user friendliness: operating them must not require specialized personnel, nor specific skills, and should be performed directly by the decision maker;
- modularity: the possibility must be granted of implementing, in a simple way, different procedures, depending on the requirements of the decision maker;
- flexibility: the capabilities of the workstation must evolve smoothly according to the evolution of the environment and of the decision maker's needs;
- integration: the workstation must provide easy ways to access all the relevant information for each decision, possibly including decisions of other decision makers in the network, and/or the relative operational outcomes; in this way LINDEN will constitute a support for taking decisions in an integrated way and for integrating decisions.

As a consequence, at the global network level, LINDEN will have an open architecture, interfacing and integrating different systems, within and between organizations.

A base for the logistic workstation of LINDEN is the existing workbench for logistics, already implemented by IBM Semea. One of the objectives of the project consists in improving its capabilities by producing new modules and features, and in providing it the possibility of exchanging information with every other possible element of the network. These activities will lead to design, implement, integrate, test, and assess prototype elements.

The modules of the logistic workstation will impact on several decision areas, involving issues such as:

Designing the logistic network

Relevant tasks in this area will be, for instance:

simulation/emulation: providing quantitative performance figures for a possible—or actual—implementation of the logistic network, corresponding to different scenarios—or evolving from the real situation—this feature must be implemented at each appropriate detail level for the relevant decision level, and must encompass the possibility of "blowing up" critical elements of the network for further analysis and validation;

optimization: providing effective ways to design an efficient and effective logistic network — or to enhance the performance of an existing network — meeting specific needs and requirements;

rough design: providing quick-and-dirty ways to take rough, qualitative decisions — such as assessing the feasibility of a proposal for a contract or a project, in order to accept or reject it on the spot — accounting for available resources and conditions, including uncertainties, estimating the order of magnitude of costs and revenues, and pointing out potentially critical aspects.

Other tasks in this area will concern the logistic impact of decisions involving other enterprise functions, such as product design, management of suppliers, production planning, marketing, ditribution, etc. always within a concurrent and integrated approach to problems. Examples are:

- identifying the most appropriate point of the logistic chain where the customization process of a product must begin;
- steering the evolution of all processes relative to a specific family of products when their structure changes, such as when new models substitute old ones, or some of them go out of production.

Monitoring the logistic network

Modules of this area will provide tools for identifying and gathering information and data on the operations of the logistic network in order to check the smooth process of all activities according to designed or projected conditions and requirements.

Controlling the logistic network

Modules of this area will operate on the data of the monitoring area in order to spot anomalies and identify, select, and implement appropriate and timely corrective actions.

Tasks in this area will provide elements for a *logistic control panel*, encompassing activities such as:

- prompting deviations from the expected behavior of operations in the logistic chain;
- diagnosing anomalies;

- identifying possible corrective actions, also providing insights to assess them;
- selecting the most appropriate control action;
- providing the best way to implement the selected control action.

Communication

In LINDEN, the logistic workstations of the different decision makers are merged into a communication network, supporting both public and contract information sharing and exchanging. Each logistic workstation will have appropriate modules implementing communication functions apt to integrate not only within LINDEN, but also in the relevant existing Information Systems and EDI systems.

2.2. Network level

The network level is where the innovative impact of LINDEN is also very relevant. It connects the workstations of the different decision makers up to the appropriate level of integration and allows them to transmit the data necessary to decisions. At this level, essential features are again:

- modularity: introducing new users in the network, with the appropriate access and integration levels, must be done in the easiest way; in this sense, the network must have the capability of evolving dynamically according to the requirements. As an extreme example, temporary partners in a logistic contract should be able to join the network — or create a new one — for the single purpose of operating the contract activities, and then to leave it, with limited installation costs.
- flexibility: users with different information systems must interact in a totally transparent way, at the desired level of integration.

These requirements raise relevant problems about communications, access levels, data privacy, standards, protocols, etc. They will be addressed from either point of view of intra-and inter-organization. At this network level, base elements for LINDEN, besides the logistic workstations, are databases and information systems containing pertinent information, and existing Electronic Data Interchange systems for communication. Special care will be devoted to identify in each case the relevant data to be accessed and forwarded to other actors, and the data privacy protection measures.

Accordingly, another objective of the project is to design the general architecture of LINDEN and to provide guidelines for implementing it, on the base of pilot installations tested by the partners.

3. PARTNERS

The proposal is driven by IBM Semea with some Academic and Research Italian partners, from Universities of Rome and Trieste, Polytechnic of Turin, ISTIEE (Institute for Study of Transportation in the Integration of European Economies) of Trieste.

However, as it was mentioned in the Introduction, the project is open to contributions from other partners. It is therefore appropriate to sketch here possible interests, motivations, and roles for these partners.

3.1. Profile of other partners

Other partners are required to collaborate in the phases of:

- identifying and assessing the needs of the logistic manager;
- analyzing, designing, and improving decision procedures;
- identifying relevant information for the decision procedures;
- analyzing, designing, and improving procedures for accessing, structuring, and using relevant information;
- deciding data accesses and data privacy protection issues, and formulating guidelines for them;
- testing and evaluating prototype modules;
- extending the above issues in order to apply them to the widest range of potential users, especially small and medium firms operating in the same or similar areas.

Although no interaction is required for the phase of prototype development, it could be helpful.

Prospective partners should fall into one of the following classes:

- **industrial partners**: large companies producing widely distributed and diversified products, controlling most elements of the logistic chain from raw material suppliers to end customers. Companies operating in areas different from those of IBM would be more advisable, such as food, chemicals, pharmaceuticals, etc.

- **logistic partners:** large logistic firms, providing international distribution/logistic services, such as shipping, handling, packaging, warehousing, transportation, etc. Besides their own needs, views, experience, and procedures, such companies should also provide insights on the concerns of small/medium size firms operating in the same area.

A sketch of possible benefits for partners

- acquiring tailored tools and solutions for their needs in logistics, especially devised and implemented according to their requirements;
- taking advantage from researches, know-how, solutions, tools, and prototypes already developed by IBM;
- investigating, assessing, and gathering opportunities from integration of the logistic and industrial partners;
- profiting from integrated fund rising projects, e.g., from European Community, such as ESPRIT;
- benefiting from the enhanced visibility within the European arena provided by concurrent activities;
- cooperating with advanced research centers for devising breakthrough solutions;
- enhancing solutions availability in logistics as an emerging business area.

Estimated costs

A tentative term for the project duration is 30 months. A rough, preliminary estimate of the resources required (in *men* x *months*) is:

Module develoment and test	260
Network design and communication	220
Pilot implementation	180
Project management and documentation	140
Total	800

4. DESCRIPTION OF THE PROJECT

4.1. Local level

The local level of the project consists of designing, implementing, and testing a prototype of the *logistic workstation*. The logistic workstation has a modular structure: it contains several *modules,* each providing a particular function to support the needs of the *logistic manager* in order to allow him to take the best decisions.

Selection of the modules
The modules to be developed in the LINDEN project have been identified on the base of a pre-analysis of the needs of a logistic manager carried out during the phase of feasibility assessment of the project proposal. They are

not intended as exhaustive, with respect to the structure of the logistic workstation, but rather as a set of convenient tools for the most relevant and immediate needs of the logistic decision maker, open to further additions and improvements. Verifying the soundness of this choice is a part of the project that will be carried out during the phase of analysis of each module. To this end, all the partners will exploit their experience in order to account not only for their enterprise requirements, but also for the requirements they perceive in different actors of their business environment. In this perspective, a special attention will be devoted to the needs of small/medium companies, that hardly have the possibility of carrying out such analyses, but represent a wide market of potential customers and users of LINDEN.

Modification of the modules
The possibility of modifying even relevant features of the modules with respect to how they are presented in the present proposal lies within the scope of the project.

Specific alternatives that should emerge will be evaluated according to their potential impact to the effectiveness of the relevant module, under the rigid constraint that they must not alter the resource level required from the partners of this project

The opportunity of providing new modules, not included in the present proposal, will also be considered. However, they are not expected to be developed within this project. The possible indication of the need for them, with a rough specification of their features, and the related motivations, are considered as a potential result of this project.

Description of the modules
The activities of the project at the local level are devoted to design, implement, integrate, test, and evaluate the modules described below. The concept of *integration* is basic throughout the whole project and bears a special importance within the context of the logistic workstation, with a significant impact upon how this part of the project has to be carried out.

a) Integration of the logistic workstation   Integration plays a threefold role for the logistic workstation:

- **internal**: all modules must smoothly interact, under the control of the decision maker, in order to provide an effective support to her/his decisions; as a consequence, modules must have total compatibility with respect to the data they use and produce. Furthermore, the possibility must be granted of assembling them within procedures, devised and implemented interactively by the decision maker in the course of taking her/his decisions; such procedures must be recorded, modified and used in the most friendly way, without requiring specialized personnel, nor specific skills;

- **enterprise**: the logistic workstation must provide an easy access to all the relevant information which is necessary to enhance the quality of the decisions of the logistic manager; in particular, it must have convenient interfaces with all the pertinent databases, information systems, and EDI systems of the decision maker's company. This requires an open structure for the architecture of the logistic workstation, in order to interface it with different systems. During the course of the project, the prototype workstations that will be provided to the partners for the evaluation phase, will be connected to their information systems in order to also assess their customization attitudes;

- **network:** the logistic workstations must also interact at the network level of LINDEN. Therefore, each of them will have appropriate modules implementing communication functions. Developing them concerns part of the activities of the network level of the project.

In the following the modules proposed for this project are described, and some relevant requirements for them are outlined. Checking and specifying such requirements will be part of the analysis phase of the development for each module.

b) Simulation module This module describes the system dynamic evolution and provides quantitative performance figures for a possible implementation of the logistic network, corresponding to a specific scenario that represents one of the alternatives among which a selection has to be performed. The possibility must be provided of using this module at each appropriate detail level, for the relevant decision level, and of "blowing up" critical parts or elements of the network for more disaggregate analyses and assessments.

c) Emulation module This module describes the system dynamic evolution and provides quantitative performance figures for a possible implementation of the logistic network, corresponding to the actual measured state of the logistic processes. It is similar to the simulation module, except for the fact that now data represent the actual situation in place of hypothetical scenarios. This difference is specifed by using the term *emulation* in place of *simulation*. Despite such a strong similarity, the development of these two modules is kept separate, since their structures are expected to be sensibly different. In fact, simulation requires a rather complex stochastic machinery, whereas emulation needs interfaces to gather data on the process states.

d) Optimization module This module provides tools to design an efficient and effective logistic network — or to enhance the performance of an existing system — meeting specific neeeds and requirements. The most relevant decision variables considered are: shipment sizes and frequencies, routing, reorder points, carrier and stock capacities, etc. Different optimization algorithms will be implemented for this module, with different features, ranging from general purpose strategies operating on specific models, up to special purpose methods devised for specific problems. The possibility of using, in an integrated way, Artificial Intelligence approaches will also be investigated for this module, and will be implemented whenever appropriate.

e) Rough design module This module provides "quick-and-dirty" ways to take rough, qualitative decisions on logistic processes, such as assessing the feasibility of a project or a contract proposal, in order to accept or reject it on the spot, accounting for available resources and conditions, including uncertainties, estimating the order of magnitude of costs and revenues and pointing out potentially critical aspects. This module could share several characteristics of the optimization module, such as the opportunity of encompassing different optimization methods and AI approaches. The substantial difference is that now a very good time performance has the highest priority. This requirement is expected to make ineffective the solutions devised for the optimization module.

f) Customization module This module addresses the problem of adapting the logistic processes in order to support the customization operations for products requiring them. These operations may have a relevant impact on several activities of the logistic network, and also involve other enterprise functions, such as product design, management of suppliers, production planning, marketing, distribution, etc. always within a concurrent and integrated approach to the problems. As an example, identifying the most appropriate point of the logistic chain where the customization process must begin is a relevant problem within this framework.

g) Product substitution module This module addresses the problem of steering the evolution of all logistic processes relative to a specific family of products when their structure changes, such as when new models replace old ones, or some of them go out of production. This module provides tools for identifying and gathering information and data on the operations of the logistic network in order to check the smooth evolution of all activities according to designed or projected conditions and requirements. Data acquired by this module could be used as inputs for the emulation module and for the control module.

h) Monitoring module   This module provides tools for identifying and gathering information and data on the operations of the logistic network in order to check the smooth evolution of all activities according to designed or projected conditions and requirements. Data acquired by this module could be used as inputs for the emulation module and for the control module.

i) Control module This module operates on the data provided by the monitoring module in order to spot anomalies in the operations of the logistic network and to identify, select, and implement effective and timely corrective actions. It constitutes the **logistic control panel**, encompassing activities such as:

- prompting deviations from the expected behavior of operations;
- diagnosing anomalies;
- identifying possible corrective actions, also providing insights to assess them;
- selecting the most appropriate control action;
- providing the best way to implement the selected control action.

Also for this module the opportunity of encompassing both optimization and AI procedures will be considered.

Development of the modules.   The project activities devoted to designing, implementing, prototyping, and testing the modules of the logistic workstation will be carried out in a modular way, performing the same sequence of phases for each module. The phases are:

1. **analysis**: the features and requirements of the module are specified on the base of the partners' experience;
2. **solution**: the most appropriate models and solution methods are identified, selected, and verified for logical soundness;
3. **implementation**: the selected models and methods are implemented in a prototype, stand-alone module;
4. **testing**: the prototype module is tested on case problems to verify its actual fitness to the relevant situation and to assess its compatibily in terms of the computational resources they require;
5. **integration**: the prototype module is merged within the structure of the prototype logistic workstation;
6. **piloting**: the prototype integrated module is tested on the field in a pilot site by one project partner against the requirements set in the analysis phase.

Clearly, two main feedback loops must exist for this sequence, starting after the testing and piloting phases if they are not passed. Milestones in the module development procedure are:

- completion of analysis;
- successful completion of testing;
- successful completion of piloting.

A typical leadtime for the module development procedure is ten months. It is shrunk to six months for the simulation module, which can exploit products already developed. It is expanded to twelve months for the more complex modules, such as the optimization and control modules.

The main risks of the module development procedure consist of the possibility of not passing the testing phase or the piloting phase. In either case, a *failure analysis* phase will be undergone, devoted to planning an iterated crash procedure. Typical leadtime for iterations are one month after testing failure and two months after piloting failure. The option of giving up the development of a module will be considered after two iterations, and will be evaluated according to the global situation of the project, taking into account the possibility of modifying the resource allocation among the workpackages and of increasing the resource level.

4.2. Network level

The logistic network is a complex environment where several actors interact in order to supply, store, ship, and distribute various commodities. The decision makers in these processes belong to different groups, departments, and companies. Nevertheless, they must collaborate in order to operate the logistic chain. Exchanging information for decisions is one of the ways in which collaboration takes place. To support these processes in the logistic network, the workstations of LINDEN are integrated into a communication network.

When different actors collaborate in an integrated way within the logistic network, we say that they are partners in a contract. Operating a contract is the way in which logistic services are exchanged between suppliers of logistics and customers of logistics.

LINDEN supports two kinds of processes related to contracts that take place in the phases, respectively, when partners are sought and when the contract is operating. Accordingly, the network level has two sublevels: public and private, dealing respectively with decisions of these two phases.

Whereas at the local level the LINDEN project aims at designing, implementing, and testing modules for decision supporting, at the network level the objectives are, in the present formulation, somehow less ambitious. They consist essentially in a feasibility study, devoted to:

- check the logical and technical consistency of this part of project;
- develop requirements for functions;
- establish guidelines for activities.

As a consequence, this part of the proposal is less specific and detailed than for the local level. Conversely, it is more open to suggestions and contributions.

Public sublevel

At the public sublevel of the network level, LINDEN supports processes devoted to setting up a contract. This is the situation when, for instance, some actors have a logistic need and seek partners able to cope with it. As an example, a producer may look for shippers, carriers, and warehouses to define the logistic network for a new product in the design phase. Alternatively, there may be a supplier of logistic services looking for clients. As an example, a traffic manager may look for clients in order to consolidate small shipments.

At the public sublevel, LINDEN is a mean allowing to circulate public domain information (requests, offers, needs, opportunities, etc.) of actors requesting or providing logistic services, in a sort of electronic advertising.

Essential features of LINDEN at the public sublevel are user friendliness and felxibility in order to allow any actor to subscribe and operate in the easiest way, using his everyday equipment (both hardware and software).

Activities that must be granted to actors at this sublevel are:

- inserting an advertisment
- asking questions, looking for answers to their needs
- establishing a communication between prospective partners
- passing to the private sublevel when the contract is stipulated, in order to manage it.

From a functional point of view, there are essentially two types of functions at the public sublevel:

- data base management
- inter-actor communication.

Both functions have modest dynamic performance requirements. A possible operational paradigm for the latter could be electronic mail, providing a way to "post and circulate advertisments" in an asynchronous way at a low integration/synchronization level.

A wide reachability of messages will be a major feature, with no, or modest, relevance of privacy issues.

Private sublevel
At the private sublevel of the network level, LINDEN supports the decisions of partners in a logistic contract using the different modules designed for the local level. The network aspect of such an activity allows supplying the appropriate information for the decisions. Here communication has higher performance requirements than at the public sublevel: the possibility of establishing two-ways point-to-point connections is required, operating in real time and allowing dialogue. Selective reachability is essential, and this raises major privacy issues, in order both to keep private communications and to guarantee the appropriate level of data privacy for each partner.

The data base features are less important now than at the public sublevel. Conversely, access to data from physical processes is now essential. Again, this raises further privacy problems and harder requirements for real time operations.

ACKNOWLEDGMENTS

The work of Walter Ukovich has been supported by CNR (National Research Council of Italy), *Progetto Finalizzato Trasporti 2*, under contract no. CO93.01906.74.

REFERENCES

1. **Carlos F. Daganzo**. *Logistics Systems Analysis*, volume 361 of *Lecture Notes in Economics and Mathematical Systems*. Springer-Verlag, Berlin, 1991.
2. **Mark S. Daskin**. Logistics: an overview of the state of the art and perspectives on future research. *Transportation Research A*, 19A(5/6):383-398, 1985.
3. **Robert L. Flood** and **M.C. Jackson**. *Creative Problem solving: Total System intervention*. Wiley, New York, 1991.
4. **Graves, A.H.G. Rinnoy Kan** and **P.H. Zipkin** (eds.). *Logistics of production and inventory*. Handbooks in Operations Research and Management Science, Vol. 4. Elsevier, Amsterdam, 1993.
5. **James C. Johnson** and **Donald F. Wood**. *Contemporary Logistics*. Macmillan, New York, fourth edition, 1990.
6. **Randolph W. Hall**. Research opportunities in logistics. *Transportation Research A*, 19A(5/6):399-402, 1985.

7. **Ari-Pekka Hameri** and **Eero Eloranta**. Case studies in logistics. In Eero Eloranta, Editor, *Advances in Production Management Systems*, pp. 195-204, Amsterdam, 1991. North Holland.
8. **Ralph L. Keeney** and **Howard Raiffa**. *Decisions with Multiple Objectives: Preferences and Value Tradeoffs*. Wiley, New York, 1976.
9. **Ralph Sims**. *Planning and Managing Industrial Logistics Systems*. Elsevier, Amsterdam, 1991.
10. **Walter Ukovich**. Modelli normativi per la logistica - Rapido viaggio nell'universo periodico, e dintorni (in italian). In: G. Di Pillo (Ed.): *Metodi di ottimizzazione per le decisioni*. Masson, 1994, pp. 381-397.

Chapter 9

AN INTEGRATED APPROACH TO DESIGN INNOVATION IN INDUSTRIAL LOGISTICS

Marco Crasnich
Consultant, Monfalcone (GO) - Italy
Camillo Lanza
Consultant, Milano - Italy
Anita Merli
IBM SEMEA, Vimercate (MI) - Italy

1. INTRODUCTION

In the last years, many industries saw a steadier increased segmentation of their production processes, while their markets gained increasing transnationality.

This emphasized the importance of logistics as the nervous system of a company, enabling it to integrate and coordinate its activities.

Increased market competition demands enhanced performances in every area of a company and especially in key areas such as logistics, where more benefits and synergies can be obtained.

Logistics could really increase companies competitiveness but too often logistic processes and activities are not designed and planned considering all their further implications; they are rather born as an outcome of the evolution of specific and sometimes temporary situations. The result of this approach is a logistic structure that may be effective but is often not efficient in terms of cost/benefit and in most case is not optimized.

For a manufacturer, logistic network design comes generally after production design and depends on the product structure and the production process. In such an approach, logistics plays a secondary role and has few chances to reduce cost and improve overall performance.

Just as a concurrent engineering approach is often used to achieve an integrated production process and product design, it is also clear that to improve enterprises' business we must have a structured and integrated design approach to logistics.

The complexity of the problem is very high in terms of relevant factors, number of data to consider, and speed of environmental evolution.

This situation involves a continuous and dynamic analysis and control of the decisions taken.

It is therefore necessary to have suitable tools for supporting and managing these activities.

In such a framework, the aim of this paper is to present some tools developed to support decision making in logistics.

The content of this paper is as follows: in the first part, the "logistic workbench approach" is presented and discussed. Then some specific modules of the workbench are illustrated, showing the tasks they can perform. A special emphasis is devoted to the "frequency module", which can derive shipping frequencies when only a discrete set of feasible frequencies is available.

The mathematical programming model, upon which the frequency model is based, and some computational experiences are also presented.

2. WHY A WORKBENCH

The old approach, based on the use of system, rely on nightly batch update to get information and modules with separate database for planning, execution, inventory, that don't share information and don't communicate is not still possible.

From the other side, the prime task of a logistic is to project, plan, and control materials processes, not to become an expert in the detailed technicalities of computer operations. Thus wherever possible, computer applications must be provided which offer the maximum productivity advantage to the user but with the minimum of personal impact in terms of learning and task restructuring.

Within this scope and thinking of the new opportunities that technology can offer, IBM manufacturing plant of Vimercate (Milan) has launched a logistic workbench project.

The goal of this project was to provide new, state of the art tools to support logistic network activities and to enhance the productivity of logistics employees by providing them with an integrated set of tools running under a common operating system on a common hardware platform.

3. WHAT IS A WORKBENCH

We use the term "workbench" in this context to indicate a set of modular software tools integrated within an open architecture to provide a professional user with a friendly, powerful, and flexible working environment.

In the experience of IBM Semea, workbenches have been built on workstations based upon personal computer hardware and OS/2. OS/2 gives the advantage of a user-friendly environment and provides tools that offer an easier access to host data, with improved methods of processing the data locally on the personal computer.

Statistical and graphical tools, text editors, and many other utility programs, as well as newly developed applications and access to host applications, may coexist and exchange data in the easy to use OS/2 environment.

All these tools can be integrated into customable environments by which the users are led through the applications that support their daily activities.

Local manipulation of data provides the possibility to change data and experiment different alternatives while preserving the integrity of host database. This is an essential feature for a structured design approach to logistics. Besides, the use of a workbench allows avoiding costly host processing taking advantage of personal computer power that, in most cases, is underutilized.

Furthermore, the workbench approach gives a tremendous increase in productivity with, at the same time, more satisfaction to the user who will certainly gain in professional skill. We may assume that the workbench represents, for logistics as well as for other types of users, a new way of doing business.

New is the concept of a menu, shown in the workstation, that drives your activity by telling you, by type of process, the applications available to support a specific task.

The workbench approach gives the users a facilitated access to complex host and PS applications. All systems are seen through the same standardized access interface; moreover, working on a personal workstation allows faster responsiveness and greater flexibility in business process execution.

To a certain extent the workbench can be considered like a host available under your office desk. For example, you could easily approach a job on inventory by accessing through your workstation an inventory profile application, obtain from it the requested data, access a logistic network simulation application, gather other type of data, then go through a navigation process and compare results with previous data. With a click of your mouse you can go on to a charting tool, still shown in the menu, obtain a graphical version of your data, tailor the graphic into a host e-mail document, and using the 'cut and paste' facility add more data and send it out.

In addition to the use of the workbench environment as a means for integrating different information and providing an easy to use tool to manage the information obtained, specific logistic modules have been prototyped.

This set of tools, integrated into a workbench-like environment, has been named Logistic WorkBench (LWB).

The modules developed so far cover the areas of Network simulation, Kanban Engineering, Network monitoring and Inventory profile analysis.

4. NETWORK SIMULATION

The Network Simulation application is a flexible simulation tool to allow the end user to easily build a model of a logistic supply network. The purpose of this application is to verify the proper dimensioning of production and transportation resources, logistic parameters, stocks and buffers, and to evaluate the performances of the network, both for an entirely new product or to support the redesign of an existing network (e.g., new sourcing, major process redesign, change in demand needs, etc.).

The model is described by defining product structure (generally limited to high dollar part numbers), product demand, and network structure. The network structure includes production and external supply nodes, transports, and ordering rules. Demand can be defined in terms of volumes plan and random variability with respect to planned volumes. Production and logistic parameters can be defined at node or part number level. Random variability of production and transportation facilities performance is supported. Parts can be ordered in a pull fashion (based upon actual consumption) or using an MRP logic (planned demand anticipated by the lead time). Only one source per part/number is supported in the first release, although a multiple sourcing model is already under study.

Data loading interfaces will allow integration with company databases to extract product structure and logistic parameters. Simulation uses the RESQ simulation package, running either on a VM host or on an AIX workstation.

Network Simulation can share data with the Inventory Profile module; the two tools can be used together to evaluate both performance and inventory impact of different network design alternatives.

The application allows to estimate expected production of each product at each node and compare it with demand, thus estimating network flexibility and performance. It also enables automatically calculating minimum buffer levels required to support a given demand variability. Results are presented in table or chart formats.

5. KANBAN ENGINEERING

The Kanban Engineering application is a simulation tool focused on an individual pair of user/supplier linked by a pull (kanban) ordering technique. With respect to the Network Simulation model, this tool offers a much simpler, faster to use model, using a very limited amount of data. In many cases a quick estimate of the proper dimensioning of a kanban area can be very valuable. Moreover, this tool offers a more sophisticated and detailed model of the pull ordering mechanisms.

The application can be used to either evaluate the service level offered by a given kanban buffer dimensioning, or determine the minimum kanban size to provide a given service level.

Also, in this case, simulation uses the RESQ simulation package and may run either locally on PC, or on a server machine (VM host or AIX workstation).

Due to the limited amount of data required, Kanban Engineering generally doesn't require interfaces to host databases.

6. INVENTORY PROFILE ANALYSIS

Another important task in the logistic engineer activity is the analysis and evaluation of financial impact related to the production and management of the products.

The inventory profile modules gives the possibility to a LWB user to evaluate the inventory profile and related turnover for a new product as well as for a current product where process reengineering needs to be performed.

The purpose of the application is to offer a fast way to quantify the results, in terms of product cost build up and delta turnover, due to a change in the parameters, costs, sources, product structure, etc.

For that reason the Profile Analysis module offers its maximum value added when used in a close feedback loop with the Network Simulation one (see Network Simulation module description).

Due to its nature and scope, the Profile Analysis doesn't replace any mainframe applications already in place to monitor and control either the inventory profile or the inventories level during the total life of a product; it rather supplements them for fast and easy what-if analysis of changes in the supply network.

Inventory profile can share data with the Network simulation module; the two tools can be used together to evaluate both performance and inventory impact of different network design alternatives.

7. NETWORK MONITORING

In addition to the design activity, it is fundamental to monitor the system to realize an effective control of the entire process and to close the manufacturing process in a design-control-redesign loop.

The Network Monitoring module provides the possibility to monitor and control the behavior of a given network.

The main objective of this module is the integration of data on orders/products/parts, already available on different applications and databases, in a unique environment, to allow physical monitoring and status tracking.

This module, integrated in the Logistic Workbench, also offers the capability to evaluate, by measuring the cycle time between two events of the production network, the effects of the decisions taken during the design phase of the manufacturing process itself.

8. SOLE: A TRANSPORTATION ENHANCEMENT TOOL

The openness of the workbench architecture makes it easy to integrate additional modules into it. Thus, a developed outside of the LWB project is now being integrated into the workbench. This module is the S.O.L.E (Sistema Ottimizzazione Logistica Esterna) application.

For this application we will give a more detailed problem and model description.

Decision on network structure in terms of vendor frequency linkage, routing, and vehicle types, depots alternatives have a great impact on transportation quality and cost. It is very difficult to reach an "optimum" network (minimum cost) and to maintain it only with completely manual based methods. Moreover, it is expansive to follow production variations by frequently manual based network redesign. For this reason, the transportation service could sometimes be underdimensioned (with the consequences of low truck loading saturation percentage).

The purpose of the SOLE transport optimizer is to rationalize transportation from and to locations supplying and receiving parts from a plant or a distribution center.

The goals of the application are: to increase the design flexibility, reducing the design time in order to be able to follow production's variations, to increase network efficiency using optimization criteria related to inventory and transportation costs, to support budgeting activities.

The transport optimization kernel contains a Mixed Integer Linear Programming (MILP) model and an heuristic algorithm. The former optimizes transportation frequencies, and the latter calculates recommended routes.

The frequencies model considers products to be collected, their quantities, inventory costs, and physical volume. The time horizon within which all goods must be collected and the unitary time section need also be defined.

Taking into account carriers availability, their capacities, and costs, the MILP algorithm selects among a set of user-defined feasible frequencies, i.e., the number of collections which are to be performed in the time horizon for a certain product. Collection intervals need not be submultiples of the horizon length.

The output of the model for each product, is the definition of the collecting frequency which minimizes transportation and inventory costs. A collecting plan for each vendor or customer over the whole horizon is also generated, defining which product is to be shipped, in which quantity, in which day, and by which carrier.

The route model defines the routes which are to be followed every day in order to perform the planned collecting and delivery, minimizing costs and distances by grouping shipments into multipoint routes.

The main input is, day by day, a list of requested shipments (to/from plant and between vendors). This list may be partially taken from the output of the frequency model, integrated with additional requests and, if necessary, modified by the user.

The route model can also be used in a what-if analysis mode to simulate different choices of one or more central warehouses. The model allows using many collecting centers at the same time, providing the opportunity of evaluating costs of different alternatives.

Consistently with the Logistic Workbench architecture, SOLE features an OS/2 Workplace Shell user interface, while the kernel has been developed on a RISC/6000 platform. The optimizer OSL has been used for the frequency problem; the routing problem has been solved with a custom-made "saving" algorithm.

8.1. FREQUENCY MODULE

This module of SOLE application is about the links between a factory and its suppliers: it is the search for the connection frequency that minimizes the sum of transportation and inventory costs.

The connection with every supplier is analyzed: the plan of movements in a given time horizon, the prices of involved part/numbers, the distance from the vendor, and the features and costs of the available carrier fleet are under examination.

Running philosophy

The user decides which is, for the supplier to be investigated, the set of linking frequencies interesting to examine: it is useless and having opposite effect to consider frequencies not belonging to the real range of acceptability.

Then the user decides the calendar of the days allowable for transportation: holidays will be cancelled from the considered horizon.

Also for the list of the carriers is good to be concise to avoid waste of CPU time.

The purpose to reach is singling out of the frequency of connection between the factory and a supplier which minimizes the sum of transportation and inventory costs.

For computing transportation costs it has been used a fare consisting of a fixed charge for load/unload operations and a fare by kilometer. Such fares vary according to the kind of used carrier. Carriers have different load capacity.

Inventory cost is defined as a rate (definable by the user) of the item cost, multiplied by the lot dimension and by the number of time sections which are the unit of measure of the percentage.

In this case we refer to a consignment situation, where inventory costs always carries by the consignor both for the parts and for the ended product. Therefore inventory cost depends only by the dimension of the moved lots and by the time lag spent between two following actions of delivery/collection.

The minimization has been performed by a model of integer programming by OSL package.

The optimization model

- Indexes

i - product index
j - frequency index
k - time section index
l - carrier index

- Variables

X_{ij} SOS variable, decision of moving product i with frequency j

Z_{ijk} 0-1 variable, decision of moving product i with frequency j in the day k

Y_{lk} integer variable, number of carriers of kind l to be used in the day k

- Elements

HORIZ   considered time horizon

SETFR(j)   set of possible frequencies

KMF   distance of the considered vendor from net center

SETGIO(k)   set of allowable days for shipping

BOUCAM(l,k) daily limit, lower and upper of availability of different carriers

DATI(l)   data about costs and capacity for every kind of carrier

PREZZUN(i)   unit price of every item

QTA(i)   quantity of each item foreseen in the horizon

DATI(i)   other data about items

RATE   inventory cost rate

- Objective function

$$\text{MIN} \quad \sum_{i=1;\ \text{NPROD}} \sum_{i=1;\ \text{NF}} \sum_{k=1;\ \text{NGF}}$$
$$[QTA(i)PREZZUN(i)HORIZ(RATE/SETFR(j)^2)Z_{ijk}] + \quad (0)$$
$$+ \sum_{k=1;\ \text{NGV}} \sum_{i=1;\ \text{TIPOCAM}} [(CK(i)\ 2KMF + CU(i))Y_{lk}]$$

- Constraints

For each product it is chosen one and only one frequency

$$\sum_{j=1;\ \text{NF}} X_{ij} = 1 \quad \forall i \quad\quad (1)$$

The number of shippings has to fit with the chosen frequency

$$\sum_{k=1;\ \text{NGV}} Z_{ijk} - SETFR(j)X_{ij} = 0 \quad \forall i, \quad \forall j \quad\quad (2)$$

To be sure that for every day of shipping it is available a suitable load capacity. Of course the way of computing the volume may change according to the packaging of each product.

$$\sum_{i=1;\ NPROD} \sum_{j=1;\ NF} \text{Volume } (QTA(i)/SETFR(j))\ Z_{ijk} -$$
$$- \sum_{i=1;\ TIPOCAM} CAP(i)\ Y_{lk} \le 0 \qquad\qquad \forall\, k \qquad\qquad (3)$$

This constraint assures that shippings are nearly equally spaced in the time, consistently with the allowable calendar.

$$\sum_{p=k;\ k-1\ +\ HORIZ/SETFR(j)} Z_{ijp} - X_{ij} \ge 0 \quad \forall\, i; \quad \forall\, j;$$
$$\forall\, k=1,\ HORIZ - HORIZ/SETFR(j) \qquad (4)$$

- Bounds

Variables Y_{lk} are bounded by their availability.

Computational experience

Computation has been carried out only on real life data, taken from planning files in an IBM manufacturing location: no "ad hoc" problem has been built for. Matrix generator has been written in AIX XL FORTRAN/6000

V2.2, and the optimizer is, of course, IBM AIX OSL/6000 V1.2. The hardware is an IBM Risc Sistem/6000 Model 530.

The characteristics of the problem are the following:

I number of items
J number of frequencies
K number of days of the time horizon
L number of kind of carrier

the dimension is:

I   sets of SOS variables, of J elements each
I*J*K 0-1 variables
L   integer variables
I*J*(1+K) + L*K   total number of integer variables

case	days of horizon	frequencies	number of products	kind of carriers
1	12	5	12	4
2	23	7	20	4
3	12	5	20	4
4	6	4	114	4
5	9	4	114	4

Dimensions of the problem cases.

case	number of constraints	number of variables	non zero elements	% density
1	649	828	4887	0.91
2	2744	3542	25967	0.26
3	1074	1348	8079	0.56
4	2743	3216	16353	0.19
5	3659	4596	26067	0.16

Dimensions of the corresponding models.

case	continuous optimum	best integer solution	gap	opt.	time
1	815831	933957	14.5	Y	1' and 51"
2	995678	1096603	10.1	Y	5' and 17"
3	527526	628152	19.0	Y	3' and 10"
4	2638170	2762232	4.7	Y	3' and 8"
5	3753410	3877474	3.3	N	8' and 49"

Gap: (best integer solution-continuous optimum)/continuous optimum in %

All variables are integer.

Above results (more details in [15]) are encouraging: they define OSL and mixed-integer programming as tools of work for real life situations, unlike Speranza-Ukovich. In their paper [16], containing many ideas developed by the frequency-module of SOLE, the MIP model is presented just as a benchmark for an heuristic algorithm, but with OSL and RISC/6000 it is proven to be a really practicable method.

8.2. THE VEHICLE ROUTING MODULE

The second module of SOLE application is devoted to solve a vehicle routing problem. A manufacturing firm has to transfer goods to and from its suppliers, daily, or in a short time horizon in a just-in-time context. The available fleet contains vehicles having different capacities and cost. The proposed criteria allow to find a solution at minimum cost in few seconds.

The proposed algorithm has been tested for about 130 points of demand, but since running time depends linearly with the number of nodes, it seems reasonable to solve greater problems with it.

The literature about vehicle routing is very wide, and this is an opportunity for interesting readings. The reason is that "vehicle routing" means delivery of newspapers and milk, school bus routing, and other many things. In fact, many problems like these are treated in Pearn [4] and Pearn-Assad [5]: but in such approach arc demands are required to be nonnegative, unlike our problem. Papers by Kulkarni-Bhave [9], Achutan-Caccetta [10] and Brodie-Waters [11] present very smart formulations of the problem, but they all require carriers with constant capacity, unlike the hypothesis. But the huge number of papers about VRP is due also to the fact that no general solution has been found yet.

Another target of the work is to provide a tool which allows a what if analysis about the possible location of a new depot or the replacement of an existing one; according to its empirical character, the alternative, which are very few, may be evaluated by direct confrontation of costs, instead of following the approach suggested by Berman and Simchi-Levy [3].

It is not possible to find an optimal solution for such problem, even if it is far from the generality of a classic formulation, being oddly constrained and unsymmetrical as any real life problem. This sentence is based upon experience. Consequently an heuristic algorithm is needed, and it has to be tailored according to the features of the particular problem.

Problem description

This is the problem of a manufacturing enterprise: the enterprise has two factories, and many suppliers. Some of these suppliers are close to each other, so that it seemed advantageous to rent a store in the neighborhoods. Three depots are considered: factory 1, factory 2, and the store.

Each depot owns a vehicle fleet: the interesting features of each vehicle are type, availability, and capacity. All routes beginning in a depot have to finish in the same depot.

The matrix of distances between every pair depot/supplier and supplier/supplier must be known.

Every day there is a list of transportations to perform:

- Raw material to deliver from depots to suppliers
- Semi-manufactured items to transfer from a supplier to another
- Finished items to collect from suppliers to depots
- A shuttle activity between depots.

Others data, as containers required for each item and their dimensions, must be known.

The target of the work is to carry out all required transfers, minimizing the number of carriers and covered kilometers. As a matter of fact there are two objectives: minimizing covered distance, and to have full trucks traveling.

A solution

These two objectives could be joined together by minimizing costs, but cost elements are so various that would generate a very complex decision tree.

So the choice of minimizing the usage of carriers (as distance and number) having satisfied the need of transportation has been considered not equivalent, but at least consistent with the target of minimizing costs.

Other choices have been taken in order to simplify this problem, and most of them suggested by its features.

The list of the transfers to execute contains mainly movements involving a depot: the direct links between vendors are a small share of the list, the movements between depots are an important share of the list and a service of shuttle between them must exist, so it looks quite obvious to consider three (or whichever) single depot problem instead of a multi depot problem.

The direct links can be joined to a depot or to another by some criterion.

Shuttling and delivering-collecting activity have different peculiarities: movements between depots have to be done every day, and they involve a large amount of material. On the contrary the vendors to visit are very often changing, and the volume of the goods to be transported changes as well.

So shuttling is solved by a there and back journey, provided that it is possible to fill properly the most suitable truck.

A there and back policy is also adopted in depot vendor connections when in both ways the truck filling is greater of a given parameter.

After these simplifications the problem is reduced to its real kernel.

It has been solved by a heuristic way. The main inspiration comes from Clarke & Wright algorithm, as presented in Golden-Magnanti-Nguyen [1] and Nelson-Nygard-Griffin-Shreve [2]; these papers, unfortunately, don't care at all about the truck capacity utilization.

The basic criterion used to build not optimal but "good" routes is saving.

Assume we need to link two points i and j with depot A; with a direct link the covered distance shall be:

$$2 (d(A,i) + d(A,j))$$

joining the points in the same route we obtain

$$d(A,i) + d(i,j) + d(j,A)$$

with a saving

$$s(i,j) = 2d(A,i) + 2d(A,j) - (d(A,i) + d(i,j) + d(j,A)) =$$
$$= d(A,i) + d(A,j) - d(i,j)$$

and the steps of the algorithm are based upon this observation.

Some papers suggest to use a modified expression of the saving

$$s(i,j) = d(A,i) + d(A,j) - GAMMA * d(i,j)$$

where GAMMA, called 'shape', is a way to tune the weight of the distances between drops.

Before beginning the seek of the greatest saving, we sort our carriers in order to use first the cheapest one. Of course it is the cheapest for unit of load provided that we succeed in filling it up. Settle the maximum number of suppliers you want to visit in a single route, and the maximum distance you want to cover.

After that
1. compute the distance savings for all the sets (pairs) of nodes (suppliers) which are in the list of the day
2. sort the savings in decreasing order
3. take the first and

if the vehicle capacity is exceeded in a node preceding the last discard the pair and go to 3;

if the vehicle capacity is exceeded in the last node, fill as possible the truck and get out;

if the vehicle capacity is not exceeded, create all the set joining to the previous one of the other nodes to visit

if maximum number of drops or maximum distance is exceeded, get out

otherwise go to 1.

When you get out, there are several things to do.

Maybe you got out without filling the "cheapest" truck, so choose between the available vehicles the smallest one which can contain your maximum load, and mark it as no longer available. Net the arranged load from the list, and write out.

Go on until the list is empty.

At the beginning, when a pair of nodes is involved, of course the saving is the same regardless of the sequence, but this is no longer true for more numerous sets. The rule followed here is that a route is created by adding one of the unsatisfied nodes at the right and at the left of the existing and feasible previous set. The saving related to a set is very easily computed from the saving of the previous one, and is due both to the new node and the extreme one.

We are aware that it is a limiting choice to add new nodes only at the extreme, but the only way to get out (although not optimally) a combinatorial problem is to reduce choices.

Another assumption of the heuristic is that each node is to be visited as less as possible. Remember that we already got rid of huge volumes from and to the depot by the there and back journeys settled in the pre-processing phase. So the load feasibility test is performed every time that a node is added for all the nodes of the route. This to be sure that all the goods waiting in the depot, or at the vendor's, to be delivered at the end or during the route, can be loaded.

But unfortunately such an overdose of common sense is not enough to solve a real, practical case.

It may happen that you have to perform a direct link between two vendors, and it is so disadvantageous that the greedy policy of the algorithm will not include it in any route. But you have to transfer goods all the same.

Another practical trouble is that we can obtain a route with a poor filling of the vehicle. This can be tuned down by changing some sort criterion when the filling rate is smaller than a given parameter.

Of course the user can decide whether to perform such route or to add the charge to the work of the day after.

The problem of measuring the performance is still open. None of the refined criteria suggested by Bodin-Rosenfield-Kydes [14] can be applied to a rough solution like this. The optimization direction are two: distance and capacity utilization, and we must be very careful in not choosing indicators more sensitive to the improvement in one direction than to the deteriorating in the other.

Comparison of costs between the found solution and a real empirical solution by a parallel run seems the only way of evaluation.

Some results

One of the basic hypothesis of this work is the adherence to practical needs. So the test has been done on real life data. The involved vendors are about 130: it is not an impressive number, but big enough to kill every branch and bound technique. The algorithm has been tested for a large set of input data. For the presented problems, the time is about 18 seconds.

Comparison of a week: the data of delivery/collecting of a week have been processed by the VR module of SOLE, obtaining some routes which have been compared with the routes really settled by the people who organize transportation professionally.

DAY	Distance "empirically" covered in km.	Distance proposed by SOLE in km.	Delta in km.	Delta %	Delta Cost %
Monday	1163	972	191	16%	7.42%
Tuesday	1172	888	284	24%	24%
Wednesday	1518	898	620	40%	9.8%
Thursday	1742	1092	650	37%	11%
Friday	1453	1200	253	17%	14%
Total	7048	5050	1998		
Average				28%	12.5%

A week is a limited test, but the advantage is evident, and expected to be confirmed in many cases.

9. PERSPECTIVES

The modules developed so far are focused on a manufacturer's point of view; nevertheless, quality, cost service, and other important parameters related to products are not just a manufacturer's concern.

Time and cost performance for the handling of raw materials and distribution of finished goods are strongly affected by performance of logistic service providers.

For this reason we believe that a breakthrough advantage may come from an extension of key concepts of the logistic workbench to all the elements on the logistic chain.

Any enhancement of the entire logistic chain involves an effective and efficient decision process. This can only be obtained through the effectiveness of decision procedures and the quality of available information.

These requirements should be pursued by:

1. integrating the elements of the logistic chain (e.g., forwarders, carriers, wholesalers, distributors, etc.) by means of convenient support to data exchange
2. developing decision support modules specifically tailored for each step of the logistic chain.

These constitute the basic features of a research proposal driven by IBM Semea with some academic and research Italian partners. The proposal is known as LINDEN (Logistic Integrated Network for Decision Enhancement) project.

In this approach LINDEN is intended to be an integrated platform providing tools for supporting the decisions of a logistic manager at each point of the logistic network, for all decision levels: strategic, tactical, and operational.

This project aims at extending to the overall logistic chain the approach already experimented on the manufacturing link of the chain, thus providing companies and partnership of the companies with an enhanced capability to use logistic integration as a competitive weapon.

REFERENCES

1. **Golden B.L.**, **Magnanti T.L.** and **Nguyen H.Q.** "Implementing Vehicle Routing" , Network, 7: pp.113-148, 1977
2. **Nelson M.D.**, **Nygard K.E.**, **Griffin J.H.** and **Shreve W.E.** "Implementation Techniques For The Vehicle Routing Problem", Comput. & Ops. Res. Vol.12, No.3: pp.273-283,1985

3. **Berman O.** and **Simchi-Levi D.** "Minisum location of a travelling salesman on simple network", European Journal of Operational Research, 36: pp.241-250, 1988

4. **Pearn W.L.** "Augment-Insert Algorithm For The Capacitated Arc Routing Problem", Computers Ops Res. Vol. 18, No. 2: pp. 189-198,1991

5. **Pearn W.L.**, **Assad A.** and **Golden B.L.** "Transforming Arc Routing Into Node Routing Problems", Computers Ops Res.. Vol. 14, No. 4: pp.285-288, 1987

6. **Van Landeghem H.R.G.** "A bi-criteria heuristic for the vehicle routing problem with time windows", European Journal of Operational Research (to appear)

7. **Dror M.** and **Levy L.** "A Vehicle Routing Improvement Algorithm Comparison of a 'Greedy' and a Matching Implementation for Inventory Routing"

8. **Dumas Y.**, **Desrosiers J.** and **Soumis F.** "The pickup and delivery problem with time windows", European Journal of Operational Research, 54: pp.7-22, 1991

9. **Kulkarni R.V.** and **Bhave P.R.** "Integer programming formulation of vehicle routing problems", European Journal of Operational Research, 20: pp.58-67, 1985

10. **Achutan N.R.** and **Caccetta L.** "Integer linear programming formulation for a vehicle routing problem", European Journal of Operational Research, 52: pp.86-89, 1991

11. **Brodie G.R.** and **Waters C.D.J.** "Integer linear programming formulation for vehicle routing problems", European Journal of Operational Research, 1987

12. **Christofides N.**, **Mingozzi A.** and **Toth P.** "The Vehicle Routing Problem", Combinatorial Optimization, Wiley, 1979

13. **Bodin L.**, **Golden B.**, **Assad A.** and **Ball M.** "The routing and scheduling of vehicles and crews - The state of the art", Computers and Operations Research, 10 - Special Issue, 1983

14. **Bodin L.**, **Rosenfield D.B.** and **Kydes A.S.** "Scheduling and Estimation Techniques for Transportation Planning", Computers and Operations Research, 8: pp.23-36, 1981

15. **Merli A.**, **Scarioni M.** "A Pure Integer Model for Optimizing Transportation Costs", EKKNEWS, 9: pp.4-6, 1993

16. **Speranza M.G.**, **Ukovich W.** "Minimizing transportation and inventory costs for several products on a single link", 1991

Chapter 10

A CONCEPTUAL MODEL FOR ONE
OF A KIND AND BATCH PRODUCTION

Asbjørn Rolstadås
University of Trondheim
The Norwegian Institute of Technology
Department of Production and Quality Engineering
Trondheim, Norway

Jan Ola Strandhagen
SINTEF Production Engineering
Trondheim, Norway

1. INTRODUCTION

The Esprit Basic Research Action no 3143 "Towards an Integrated Theory for Design, Production and Production Management of Complex, One of a Kind Products in the Factory of the Future" was concerned with a consistent theory for one of a kind production (OKP). The ultimate goal of the research was to obtain a designer's workbench for the development of CIM in production systems. Such a designer's workbench required a method to describe operations in a production system. These operations included product design, tendering, logistics, quality control, etc.

The research action consisted of 5 technical work packages. In the first package, a current theory for describing operations in production systems was examined and described. In the second package a unified description model was developed and referred to as a conceptual model. In the third package existing workbenches for implementation were studied. The fourth work package aimed at operationalization of the model and the fifth involved demonstration by application in an industrial testbed.

This paper describes some of the results obtained in work package 2. However, it focuses on the work flow view and leaves out comprehensive parts concerning resources and organization/decisions.

The research consortium comprises:
- Technical University of Eindhoven, The Netherlands
- Bremen Institute of Technology, Germany
- GRAI Laboratory, University of Bordeaux, France
- CIM Research Unit, University College, Galway, Ireland

- Technical University of Denmark, Denmark
- Helsinki University of Technology, Finland
- SINTEF/University of Trondheim, Norway

The contributions from all these partners are fully acknowledged.

This work is also an attempt to work further in the possible operatization of such a designer's workbench, still within the work flow view. It focuses also on from where and how the data input for such a conceptual model should be captured. And something is said about possible use of such a model in a more operational setting.

2. ONE OF A KIND AND BATCH PRODUCTION

Production may be classified in many ways. However, two different ways are most commonly used, i.e., dependent on:

- Batch size
- Order release

Three intervals of batch sizes are usually distinguished:

- 1 (One of a kind production)
- A finite number (Batch production)
- Infinite number (Continuous production)

Order release may be split into two, dependent on whether the company will deliver from stock or will have to produce after the order is placed:

- Order production
- Stock production

One of a kind production (OKP) is always order production, and mass production is always in principle stock production. Batch production may be both.

Batch production has many similarities and therefore many of the same challenges as one of a kind production. This paper deals with batch production when it can be classified as order production.

In many cases it is convenient to regard one of a kind production as consisting of a number of operations. At an aggregate level, the following operations might typically apply:

- Design
- Process planning
- Component production
- Assembly

The first two are frequently referred to as engineering, and the latter two as production. Engineering and production comprise both processes of major interest.

Including the management type of operations, the manufacturing company can be regarded as consisting of three interdependent processes. This is illustrated in Figure 1. A process is defined as a set of related operations performed on, or in connection with a flow of concrete or abstract items.

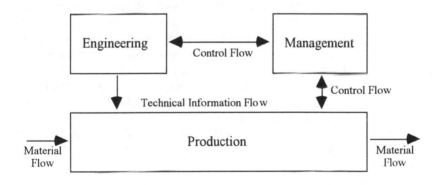

Figure 1. Processes in an OKP system.

Production is connected to a flow of materials. The purpose of production is to transform raw materials into finished products.

Engineering is connected to a flow of technical information, usually represented as drawings or other documents. The purpose of engineering is to provide technical specifications on what products to produce and how to produce the products.

Management is connected to a flow of operational information, usually represented in the form of work orders, planning, or status documents. The purpose of management is to release and monitor work orders for production and engineering.

In OKP there is always considerable engineering involved. The close connection between engineering and production and their overlap, creates many of the challenging problems in this type of production. Three different types of engineering may be distinguished:

- Engineer from scratch
- Engineer from components
- Engineer from products

An oil platform may be an example of the first type, a power station of the second type, and an aircraft of the third type.

3. MODELS

A conceptual model may be used to assist in design/redesign or testing a production system. In applying the model, some information is given as input. This is basically the products to be delivered and the resources that the company has available. Both may be inadequately, insufficiently, or only partly defined. They may all be changed or modified in the process of designing or redesigning the manufacturing system.

The company exists to satisfy a requirement for products from its customers. This demand pattern is also assumed to be one of the basic input data of the manufacturing system.

Finally, the company may operate by different working hours on different resources. The operating hours are another basic input data, of the same type as demand.

All these input data can be categorized into either *design choices (DCs)* or *experimental input*.

In designing/redesigning the system, the various structural design choices are changed to demonstrate different performance. The performance is measured by *performance indicators (PIs)*. In testing the system the structural design choices are kept fixed, while the operational design choices are changed and applied to measure different performances. Testing is to check how the structural system responds to different operational conditions.

In any application of the model some design choices and input parameters remain fixed, while others are varied. The set of fixed design choices and input parameters for a specific application is referred to as the *frame conditions* under which the model is used.

In summary, the use of the model is visualized in Figure 2.

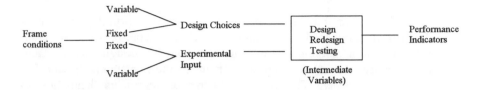

Figure 2. Input/output in operation of a model of an OKP system.

For reasons of simplification, and to reflect major different design objectives, it is convenient to consider the model from different views. In this context, three views have been selected:

- Work flow view, defining and describing the work flow through the production system.
- Resource view, defining and describing human and physical resources in the system, and how these resources respond to different input and system designs.
- Organizations/decisional view, defining decisions and decisional structure in the total organization of the manufacturing system.

These views are integrated into a conceptual model, or more correctly a conceptual framework, that ties these fragmented theories together.

The conceptual framework has to distinguish clearly between "theory" which has to do with "to describe", i.e., analysis of a model and "design" which has to do with "to make", i.e., synthesis, analysis and optimization. A terminology has been proposed to deal with these two "worlds" as shown in Table 1.

The theoretical and the design framework exist at two different levels:

a) The level of reference models
b) The level of particular models.

Reference models represent available theory. These models link general design choices (DCs) via intermediate variables (IVs) to general performance indicators (PIs). Particular models represent an abstract example of a particular case or an existing or conceived real life situation. These models show particular design choices which have been made in a particular case. They enable the designer to specify alternatives for these design choices, and to compute the consequences of these alternatives within a limited domain of knowledge.

The reference models can be split on two sub levels, A1 and A2. In the theoretical framework these levels represent:

A1 - The primitive system, i.e., the various components of the system
A2 - The constraints, i.e., how the system components are connected together.

In the design framework the counterpart to the primitive system is denoted an entity model. An entity model thus represents a description scheme for a particular real life situation by design choices and performance indicators. The counterpart to the constraints is called a *relationship model*. The relationship model represents heuristics to interconnect design choices, performance indicators, and intermediate/ independent variables of the entity model.

The components of a design reference model are described in Figure 3. The idea is that design choices (DC) are changed. The effect on the OKP system is measured by the performance indicators (PI). The relationship model defines the relationship between DCs and Pis.

Table 1. Terminology of conceptual model.

Theoretical framework		Design framework
Conceptual Reference Model represents available theory, e.g., Walrasian model	A	**Design Reference Model** represents the final "product" of project, e.g., the workbench
Model of Primitive System represents forms of data for elements	A1	**Entity Model** represents description scheme for a particular real life situation by design choices and performance indicators
Model of Constraints/Topology	A2	**Relationship Model** represents heuristics to guide synthesis. Cover "all" design knowledge about OKP. The variables in the model are DC, PI or IV and causal relationships between the variables. Thus the model increases the designer' awareness of variables which play a role in designing/ redesigning particular OKP systems
Particular Model represents an abstract examples of a particular case	B	**Particular Model** represents an existing or conceived real-life situation. Shows particular DC which have been made in a particular case

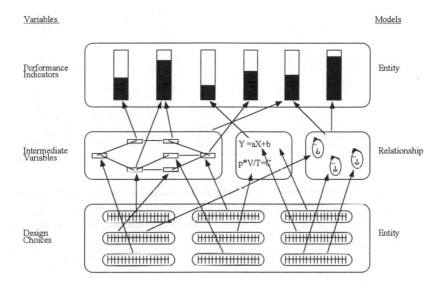

Variables   Models

Figure 3. Components of a design reference model.

4. DATA STRUCTURES

In accordance with the Walrasian model, there are two basic data structures:

- P-graph associated with the products to be produced
- R-graph associated with the resources used for production

Both graphs may be represented by hierarchical structures showing bill of material and operations of the products and the groups of resources.

The P-graph as described here, represents the primary flow in the production process. In fact, it defines the work to be done. The purpose of the engineering process is to establish the P-graph as defined here. In fact, the engineering process can be regarded as a production process, however, with a different primary flow. In this case a similar graph may be used to describe the work flow of the engineering process. Again the same is also valid for the management process.

The P-graph represents operations to be performed and defines the requirement for resources. The R-graph defines in a similar way the supply of resources, however, still independent of time.

Requirement and availability of resources are defined by two parameters:

- Capacity
- Capability

The capacity indicates how much of the resources is required or available. This is a quantitative parameter. The capability defines more precisely the type of resource needed or available. Dependent on the detailing level, the capability will be organized in a hierarchical graph structure.

The interaction between the P-graph and R-graph is indicated in Figure 4. Operation P1122 is allocated on machine R1122. Note that part P1121 is allocated on machine group R113. Such an allocation postpones a decision on details until a later stage. It indicates alternative machines possible for the manufacturing of the part (i.e., all machines in the group).

In an actual situation a demand for products is defined and the availability of resources is given in form of working hours. This means that the P-graph will define the actual requirement of components, subassemblies, etc. and, subsequently, operations in a time phase manner, i.e., the requirements will be defined within predefined time intervals. Having multiplied the requirement figures of the P-graph with the demand, time phased it, and aggregated requirements for the same operations in the same time interval, an activity is established.

Quite parallel, the inclusion of the time dimensions to the resources is established from knowing their basic capacity (units per time unit) and the working hours. These resources are allocated on the activities, i.e., an element of the R-graph is allocated to an element of the P-graph in a given time interval. This process is usually referred to as scheduling and loading. This connection represents information on:

- Which task to perform
- What quantity
- Which resource
- When and how much (of the resource).

This is the type of information we refer to as a job order.

Consequently, when P- and R-graphs are connected, and delivery time and quantity are supplied, the job orders are created.

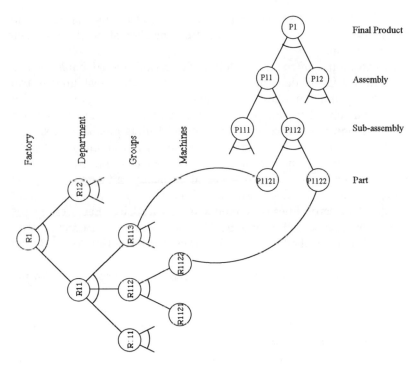

Figure 4. From Walras to ordered resources and products.

5. DATA SOURCES

To be able to build conceptual models of one of a kind and batch production systems, real data must be captured and put into the model. The sources for this data are several, and each of them must be considered carefully.

In principle the Resource-Graph data sources are more stable in time. On the other hand they are less structured and not easy to access. These sources may be:

- Machine descriptions
- Organizational charts
- Quality Assurance manuals
- Knowledge and experience of capabilities not in written form.

In most cases a Resource graph must be created more or less from scratch based on these sources. There are a number of challenges when creating the graph.

The capabilities often do not follow the organizational borders. In a group oriented layout there may exist exactly the same capabilities in two geographically different places on the floor. But for one category or class of products only one of these capabilities may be used. This must be handled by identification capabilities with, for example, drilling product type x.

Today there are ways to automatically import such data into a conceptual model. In traditional numerical models of manufacturing systems, like simulation models, this data is modeled manually, although the actual feeding of data may be automated. In such models, based on discrete event simulation, the capabilities of resources are not modeled. Each resource is identified by a name and a capacity. The actual coupling of products with resources is done explicitly, and at the lowest level referring to the P- and R- graph.

For the Product Graph, the data sources are at least the following sources:

- Bill of Materials (BOM)
- Bill of Operations (BOO)
- Drawings, technical specifications
- Process Plans
- Customer Requirements

Although on different levels of detail, and on not integrated data formats, most of these sources are easier to access. The hierarchy of a Bill of Material for a product is exactly the structure of the P-graph.

What is not explicitly found in the data sources is the information about the freedom to use alternative resources and processes, to make the products according to the specifications.

5. 1. Different phases of product development

It is necessary, in order to create a particular model, to fill in some data in the P- and R- graph. Should this data be captured at one instant of time ? Or should it be an average over some periods of time ?

In order to have as valid a model as possible, the data should be as actual as possible. But if the situation at one instant of time is very special, the data and hence, model is not typical.

On the other hand, a model based on historical data will level out any such peaks. But it might give a model which looks more to the past than into the future.

A major challenge is that at a certain instant of time, the different products are at very different stages in the production phase, namely in either:

- Request
- Quoting
- Design
- Engineering
- Production
- Assembly
- Delivery.

The amount of data for one product increases as the product goes through these processes. This is the process of giving the Product Graph more and more specified levels and exact information on each node. In the engineering and production phases the P-graph and R-graph is coupled.

This makes it necessary that a model can operate with a different information detail level of products and resources.

6. SIMULATION WITH MODELS

Using the model for assisting decisions can be described as a simulation of the models. Simulation is a term used in various contexts, and may have a different meaning in these different contexts. There are several issues that should be identified before deciding on where and when a simulation of these models is possible.

Simulation can be defined as using a software program to represent the models, and making these models run on a computer over a simulated period of time.

Three aspects are of particular interest when applying simulation in this context:

- What types of decisions are supported by simulation
- Whether to have particular models or not
- How should data values be represented

The next sections will comment on these aspects.

6.1 Types of decisions: use of simulation

There are several purposes or types of decisions where such a model may be used. These can be categorized in three:

- Design/redesign of production system
- Resource management
- Operational decisions (planning).

The accuracy, detail, and degree of particularity needed for the input for the models varies according to this use.

Simulation has traditionally been mainly applied in two areas:

- Shop floor layout and capacity decisions
- Control and dispatching (sequencing) rules

Common for both these areas is that the decision can be based on static, and therefore accurate, data. These data are somewhat randomized by using statistical distribution. These decisions are not taken daily (the decisions concerning, for instance, a scheduling rule, not a specific schedule), and the modeler and user are allowed to spend time (days and weeks) on modeling and data collection.

To apply simulation of models to assist decisions in operational issues like scheduling, would require automated (computer integrated) data collection and modeling. In addition, functions for introducing uncertainty and randomness must be present. Commerical simulation software tools still have very limited functionality in these areas.

Applying simulation models in design and redesign of production systems (today often referred to as business process reengineering) requires somewhat different additional functionality. These decisions must be taken with very incomplete information about the products. Information about some products (existing orders) are complete, as they already are being produced. While others are extremely "unknown", as they are still in quoting or engineering phase.

Redesign of a production system means redesign of both the engineering, manufacturing, and assembly functions or processes. This means that it is satisfactory only to be able to model the different process steps of products. It must be possible to model the requirements of a product. In commercially available simulation this can only be done with some simple "black box" logic, and this is not at all satisfactory.

6.2 Validation of particular models

The validation of simulation models is an extremely important topic. Validating a general model with particular qualitative data is done by looking for contradictions in the relations that makes up the model.

A particular model means a mathematical model describing a specific company. This means that the data input is collected from a company. A model requires a substantial amount of data. For a company producing one-of-a-kind products these data are simply not existing, as the future products are not designed and engineered yet.

A commonly used defintion of verification and validation is the following:

Verification - To check whether the computer simulation program and the model do what it is intended to do

Validation   - To check whether the simulation model is a good (enough) description of the real system.

A completely valid model is not possible to achieve due to the following three facts:

- A one-to-one relation between all elements of the real system being modeled, and the simulation model is not possible
- Models often picture systems, or versions of systems, that do not exist
- The use of uncertainty (statistical distribution) in the models, makes the probability that the same sequence of events should occur in reality equal to zero.

From this, the conclusion is that it is very difficult to create such models for operative use, as the uncertainty in the results from these models is large.

7. CONCLUSIONS

A conceptual model to design/redesign or test a one-of-a-kind or a batch production system, can be based on a generalization of Walras economical model, identifying two basic data structures, products and resources. These are generalized in a graph structure, P- and R- graphs. These graphs represent in data structures the required (P-graph), and available (R-graph) capacity and capability of a manufacturing system.

The Esprit Basic Research Action no. 3143 "Towards an Integrated Theory for Design, Production and Production Management of Complex, One of a Kind Products in the Factory of the Future" was concerned with a consistent theory for one of a kind production (OKP).

The work resulted in a conceptual model of Design Choices, Intermediate Variables and Performance Indicators connected by a set of relationships.

The next step is an identification of the data structures necessary for the P-, R- and D-graph, making a simulation of the management of a manufacturing system possible.

REFERENCES

1. **Falster P., Rolstadås A., Wortmann H.**: FOF Production Theory. Work Package 2, Design of the Conceptual Model, January 1991.
2. **Rolstadås, A.**: Engineering of Oil/Gas. SINTEF, Trondheim, 1990.
3. **Wortmann, J.C.**: Towards an Integrated Theory for Design, Production and Production Management of Complex, One of a Kind Products in Factory of the Future. In: Commission of the European Communities (ed.), ESPRIT'89, Proc. 6th Annual Esprit Conf. (Kluwer Academic Publishers, Dordrecht, 1989), pp. 1089-1099.
4. **Rolstadås, A.**: A Conceptual Reference Model Seen from the Functional View. Esprit Basic Research Action No. 3143, FOF/SINTEF/2-16, Aug. 1990.
5. **Takala, T.**: Design Theory for the Factory of the Future (FOF ESPRIT BRA 3143, Helsinki University of Technology, Dec. 1989).
6. **Falster, P.**: The Conceptual Model and Related Topics. (FOF ESPRIT BRA 3143, Technical University of Denmark, Dec. 1989).
7. **Marcotte, F.**: FOF: Organizational/Decisional View. FOF/GRAI/200, June 1990.
8. **Rolstadås, A.**: Structuring Production Planning Systems for Computer Applications. APMS 1987, Elsevier, 1987.
9. **Kwikkers, R.**: The Role of Simulation Models in the Conceptual Model. FOR/MGMT/173, June 26, 1990.
10. **Wortmann, H.**: Towards One of a Kind Production: The Future of European Industry. ESPRIT Project 3143. APMS'90, Helsinki, Aug. 1990.
11. **Kreutzer W.**, "System Simulation Programming Styles and Languages", Addison-Wesley, London, 1986.
12. **Law A.M.**, "Simulation Series; Introduction to Simulation", Industrial Engineering, May 1986.
13. **Birtwistle G. M.**, "Discrete Event Modelling on SIMULA", 1st edition, Macmillan Education LTD, London, 1979.
14. **Mitrani I.**, "Simulation techniques for discrete event systems", Cambridge University Press, Great Britain, 1982.
15. **Naylor T.H.**, **Balintfy, J.L.**, **Burdick, D.S.**, and **Kong Chu**, "Computer Simulation Techniques", Wiley, New York, 1966.

16. **Philips D.T.**, **Ravindran, A.**, **Solberg, J.**, "Operations Research; Principles and Practice", John Wiley & Sons, Inc., 1976.
17. **Browne J.**, **Harhen J.**, **Shivnan J.**, "Production Management Systems", Addison-Wesley, Galway, Ireland, 1988.

PART 4

INFORMATION TECHNOLOGIES FOR MANAGING
AND PROMOTING INNOVATIONS IN LOGISTICS

Chapter 11

MANAGING THE USE OF INFORMATION TECHNOLOGY TO PROMOTE PROCESS INNOVATION

Edison Tse

Department of Engineering-Economic Systems and Operations Research
Stanford University, Stanford - USA

1. INTRODUCTION

Innovation is considered to be the major source to derive competitive advantage in today's highly competitive market. Innovation is not a decision making process. We cannot derive the "optimal decision" that will maximize innovation. Innovation comes from focusing our attention in finding a creative way to fulfill the customers' recognized needs at a lower cost and a higher value, or fulfill the customers' latent needs. We usually distinguish two types of innovation as process and product innovation. In this paper we shall only discuss process innovation.

Most people perceive process innovation as the creative design and development of a new and much better production process that will produce an existing product faster, with better quality and/or at a lower cost as compared to the current production process. The most noted example is minimills which allow economic production of small batches of steel whereas the older steel mills require production of large batches in order to be economical. The new production process also gives a better product that allows producers to be a lot more economically responsive to customers' different needs . However, process innovation does not have to come from then production process alone, it can come from any of the activities in the order-production-distribution cycle. An electronic order system which provides a direct connection between a manufacturer and its vendors is an innovation that completely changes the ordering process. Just-in-time (JIT), initiated by Japanese manufacturers, is a process innovation that changes the practice in inventory control and management. The Taguchi Method is an innovation that changes the whole concept of quality control. While most PC manufacturers distribute their product via conventional computer dealers, Dell created the direct mail ordering distribution channel.

The common theme in the above examples is that a process innovation which results in drastic changes, in particular, links within the order-production-distribution (OPD) chain, allows the company to change the rule of the game in its favor in the dynamic competitive environment. The first issue is which link within the OPD chain, if changed appropriately, would allow the company to create a new competitive position. The next issue would be finding a creative way to change and improve such a link. The final issue is the successful implementation of the changes. Therefore, process innovation is very information intensive: a lot of searching, problem solving and feedback adjustment activities. Intuitively, the advancement in information technology should increase the rate of process innovation. The issue is how to manage and utilize these advances to promote process innovation. To get a grip on this issue, one must first understand the dynamics of process innovation and examine how advanced information technology can influence such dynamics.

Most of the case studies in process innovation focus at input-output status: what is the status before and after the innovation [1–3]. One can draw from each case why the innovation took place, but can draw very little conclusion on why it took place at a specific company and what triggered the innovation. Drucker [4] pointed out some of the sources for innovation but did not provide a systematic method to search for innovation opportunity. He also did not discuss the key factor for successful exploitation of innovation opportunity. There were also discussions about the importance of organization structure, incentive, and management practice that would promote process innovation (e.g., [5]). However, the literature did not provide a model that can provide insights to the dynamics of process innovation. We shall first model the dynamics of process innovation in terms of three subprocesses: a search process, a solution process, and a change management process. Based on the dynamic model of process innovation, we develop a new perspective on how to manage the use of information technology to promote process innovation.

Before discussing the dynamics of process innovation, a personal case in which I was the decision maker responsible for turning around a small business enterprise will be discussed in the next Section. In this discussion, I articulate the detailed reasoning process that I went through which led to the identification of "small" and "large" innovation opportunities. At the time of writing, the small innovation opportunity turned around the company's competitive position but the large innovation opportunity had not yet been exploited. This case study provides an introduction to the three subprocesses in process innovation.

2. A PERSONAL CASE EXAMPLE

In the mid-1980s, I was involved in turning around a pre-inked stamp business my family owned in Hong Kong. Pre-inked stamps are chemical related products which are similar to rubber stamps that many offices use for inspections, order receipts, etc. The basic difference is that instead of using rubber it uses a new material with the ink "trapped" inside.

Therefore, using a pre-inked stamp is like using an ordinary rubber stamp except that there is no need for a stamp pad. Its main advantages are ease of use and much better print quality. From an economic point of view, the pre-inked stamp has a lower product life cycle cost than a regular rubber stamp even though the unit price of the pre-ink stamp is higher. There are two major technologies for pre-inked stamps, one technology is more suitable for making stock stamps. These are ready made stamps that we find in most stationary stores with imprints like "Air Mail", "Priority", "Confidential", etc. The major player in this market is a manufacturing company in Japan. Another technology is more suitable for making customized stamps like personal addresses, company logos and addresses, special messages, signature, etc. This technology produces a premix which is the key input material to make a pre-ink stamp. The premix is made by mixing certain chemical ingredients and ink dye to form a gel like material. To date, there are only a few specialty chemical companies in the U.S. that produce the premix. Pre-ink stamp was, and still is, a niche market having a substantially smaller market share compared to the rubber stamps.

A customized pre-inked stamp is made by the following processes. The first is the editing process where the received customized orders are laid out using a word processor on sheets, each sheet contains many stamps. The second step is to make a plate just as we would do in the printing process. The third step is to pour the premix onto the plate and heat it in an oven under pressure and temperature. During the "cooking" process, the premix solidifies and forms a rubber-like slab that has ink trapped into its pores. The ink will only come out if the slab is pressed. The slab is then let to cool back to room temperature and each stamp is cut out from the slab and glued onto a stamp mount. A specially designed mount with a fine adjustment mechanism is needed to control the flow of ink when the mount is being "stamped" onto a piece of paper.

At that time, the major supply of premix was controlled by a U.S. company. Secondary sources were available, but their premix quality was slightly inferior. Before I came on the scene, our major competitor was the sole agent for the major U.S. premix supplier in Hong Kong, and we were getting our premix supply from a second source with a slightly inferior quality. This put our company in a weaker position to attract salespeople. My company had a better mount design and our competitor had a better quality premix. However, none of these was the competitive factor: the customers did not care too much about the look, and the customers could hardly tell the difference in quality. It turned out that time to delivery was the major competitive factor. Our average time was about six days, and our competitor pushed for about four days. This gave our product a slightly weaker competitive position. However, we were still the strong second in the Hong Kong market.

At about the same time, our premix supplier was bought out by our competitor's supplier. Even though our supply contract was not affected by the acquisition, it gave our competitor an angle. It launched an attack on our sales channels instead of on our customers. By spreading a rumor that my company would soon be out of business because our supplier was bought out, and coupled this with the fact that our response time was longer, it managed to hire away 30% of our salespeople. In a week, our sales dropped more than 30%. The battleground was switched from customers to salespeople. We had to instill confidence in the salespeople in order to stop a mass exodus. I went back to Hong Kong in the summer to address the emergency situation.

One of the major concerns of our sales force was that our time to delivery was longer than our competitor. This is an attribute which can easily be seen and felt by the salespeople and customers. Therefore, the first order of business was to focus on reducing the time to delivery. The first thing I did was to study the OPD chain, but I could not find the key factor to substantially improve our delivery time. However, the fact that our competitor could respond and deliver in a shorter time bothered me. I met with the factory manager and asked him how he saw the problem. He expressed an opinion that there was no problem. He claimed it was impossible for our competitor to fill orders and deliver two days earlier than we could. Probably, our sales force was exaggerating and using it as an excuse for declining sales. Yes, our premix was slightly inferior in quality but he was already doing his best. Since there was nothing he could do to improve the situation, I immediately formed an hypothesis for the key problem: it was the manager. I spent a few days confirming my hypothesis, and replaced him with my brother as the new factory manager.

My brother and I immediately started to improve every process within the OPD cycle. We discovered that many of the plant floor workers did not know why they were doing certain things. My brother, who had experience and know-how for the whole process, explained to them their responsibilities and how their activities fit into the overall process. The work flow process was reengineered, suggestions from the workers were listened to and incorporated, if appropriate, investments in new equipment were made to support the new process. Within a couple of months, the response time improved but we still could not match our competitor's delivery time. However, the workers were more knowledgeable about their tasks: how their performance could impact on the overall performance and whether they had "pushed" the existing process to its limit. It was then discovered that the product quality issue and the rework procedure imposed a limit on the response time. Since most customer orders had multiple stamps, if one of them needed a rework because it did not pass the quality standard, the whole order would be delayed. We turned our attention to higher yield rate in one pass production.

Right about the same time, the chief chemical engineer of our supplier lost his job because of the acquisition of his company by the major premix supply company. Knowing that we were one of the major users of premix, he called me one day and offered to sell us premix gel. I immediately worked out a proposal on how we could make certain financial and partnership arrangements that would allow us to manufacture premix in Hong Kong and market it to all Southeast Asia and China. With help from the chemist, we started building knowledge in gelchemistry which allowed us to figure out the source of the quality problem and solve it. We then redesigned our OPD based on our solution. Almost immediately, our delivery time was reduced from six to four days. The salespeople also noted the improvement in our product quality. The fact that we could manufacture our own supply source and deliver quality products competitively allowed us to stabilize the sales force. Since then, we steadily regained our market share.

With the existing business stabilized, I started asking the question why pre-inked stamps did not grow to take over the rubber stamp market ? What is the key factor that limits the penetration of the pre-inked products into the rubber stamp market? The price of the pre-inked stamp is higher than the rubber stamp, but the product life cycle cost is lower. So price does not seem to be a big factor. This was confirmed by our salespeople. I put the same question to my brother, the chemist, and the sales force. One speculation was that the pre-inked stamp has a longer delivery time as compared to the rubber stamp. So for many people who need a stamp in a hurry, a rubber stamp is the only answer. If this is the reason, then the next question would be what causes the difference in delivery time. In Hong Kong, rubber stamps are handmade and there are a lot of small rubber stamp makers. They can respond to customer needs very quickly. With the current production technology, production of a small batch of stamp orders is not economical. Even though, theoretically, a pre-inked stamp manufacturer can respond to a customer's order just as fast as a rubber stamp maker, it is not economically viable. Moreover, the manufacturing of pre-inked stamps is somewhat complicated, messy, and requires more sophisticated labor skills, whereas manufacturing rubber stamps is simple and easy to do. These two factors resulted in an industry structure where there are only a few large pre-inked stamp manufactures taking orders from a network of retail stamp shops. The logistic of OPD incurs additional response time for customers' orders. This creates a "negative feedback" in the pre-inked stamp market that limits its growth. Thus, the key factor that would propel further growth in the pre-inked stamp business would be to develop a new production process which is easy to use and economical in producing small batches of stamps.

The chemist's research focus was directed at developing a new process that would allow people, with very little technical skill, to make small batches of pre-inked stamps easily, quickly, and economically. At the time of writing this paper, the chemist has developed a very creative new production process that satisfies the stated objective. A prototype has been developed and tried out successfully. The next issue is how to exploit such an innovation to achieve financial success.

3. A DYNAMIC MODEL OF PROCESS INNOVATION

What I learned from my own case is that even though the end result of a process innovation is the modification of links in the OPD process, the driving force behind it comes from continuously examining how one can compete in a dynamic competitive environment by focusing on what one can do in order to change the "rules of the game". The dynamic model of process innovation consists of three subprocesses: find out what rule to change, find an innovative solution that will implement the change, and build up a new organization that can scale up the innovative solution to achieve a new competitive position. We shall model each of these subprocesses in the following subsections.

3.1 Dynamic Search for Process Innovation Opportunity

In order to change the rules of the game, one needs to challenge the established structure and practices. What are the assumptions that lead to these established structures and practices? If these assumptions are considered as conventional wisdom in the industry, then anyone who can successfully challenge these assumptions can achieve a new competitive position that no one has thought much about. This provides an opportunity for innovation. Therefore we focus our search in three directions. The first direction is to focus at the internal cost structure and uncover any possibility of changes that can shift the cost structure. The second direction is to focus at the existing industry structure. The third direction is to focus on common industry practices.

Consider a company which has an established process producing a well specified product that fulfills its customer's needs. Once the product is specified, the attributes of concern to the customers are quality, order response time, and price. Let (t,q) be a pair representing order response time and quality. One can conceptually define a "feasible" region $\Phi(c)$, which represents the set of (t, q) pair that can be achieved by a selecting a certain set of process parameters and a specific process flow that give a production cost c, and the "frontier curve" $F(c)$, which represents the best tradeoff the company can make between the order response time and quality by varying the process parameters and process flow with production cost c

(see Figure 1). As the cost c increases to c', the corresponding frontier curve F(c') will shift up. With the given production process and supply chain structure, we have conceptually a family of $\{F(c_i)\}$ as the cost varies. As c_i becomes very large, we have the limiting frontier curve F* (Figure 2).

The family of frontier curves represents the competitive profile of the company (examples of frontier curves, see [6], [8]). This family of curves is determined by the production process and the OPD chain structure that the company is adopting. By changing the production process and/or certain links in the OPD chain structure, one can change the family of frontier curves and thus change the competitive profile of the company. Conceptually, one can search for a process innovation opportunity by examining what changes in the production process, or the OPD chain, will bring about a dramatic shift in the family of frontier curves in the "right" direction to serve the targeted customer group. For example, in Figure 1, if $\{F(c)\}$ is shifted to $\{F'(c)\}$, then the company is more competitive to the customer groups that the company is currently selling to. Whereas, if $\{F(c)\}$ is shifted to $\{F''(c)\}$ or $\{F'''(c)\}$, then the company may become competitive to a new group of customers that it is not selling to currently because these customers require a delivery time or quality level that the current structure cannot meet.

Unfortunately, the analytical form of $\{F(c)\}$, and how it depends on the production process and OPD chain are not known to the company. Although we may know the direction we want to shift $\{F(c)\}$ to, we cannot determine analytically what to change in order to achieve the shift that we desire. However, to uncover what may allow us to shift the frontier curves, a search approach based on continuous improvement can be applied. The idea is that if the company is operating at an interior point of the feasible region, then it can find improvements that can locally improve quality and/or reduce delivery time without increasing the cost. Whereas if the company is operating at a point on the frontier curve, then it would discover that certain limitations disallow further improvement in both attributes without increasing the production cost. An opportunity for process innovation is to focus at removing such limitations which would shift $\{F(c)\}$.

When the competition is not that severe, a company can be profitable even though it operates in an interior point of the feasible region. Therefore in "good times", many companies having a satisfying behavior [7] will continue operating in an interior point of the feasible region. These companies will not know how to drastically improve when competition pressure builds up. This was the case that I found with my former plant manager. When the competitive pressure mounted, I had to decisively remove him so that we might learn what to do to regain the competitive edge.

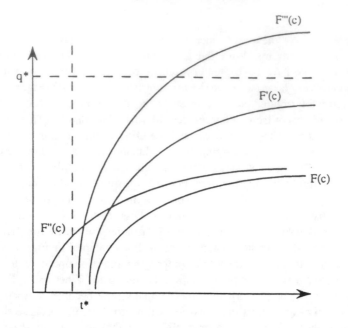

Figure 1. Family of Frontiers Curves.

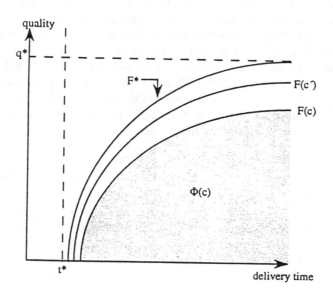

Figure 2. Shifting of F(c) to create a more competitive profile.

A similar situation occurred in the U.S. steel industry. While new foreign competitors entered the market with a new process, the U.S. steel companies high executives did not believe that such a new process would wipe out their business [4]. Operating too long in the interior point, the company did not realize the implications of the new process.

When a company is operating on the frontier curve, it is pushing the production process and the existing OPD process to its physical limit. The production team has to be very knowledgeable and skillful in carrying out the precise orchestration. Whereas, if a company is operating in the interior point, there is more room for error and, therefore, a less knowledgeable and skillful team can provide reasonable performance as expected. Therefore, in general, a manager finds that it is much less risky to operate at the interior point than on the frontier point. To reduce the risk, knowledge and skills in operating the OPD cycle are essential. To accumulate such knowledge and skills, the management needs to cultivate the spirit of relentless pursuit of excellence that would "drive" it is operating point from an interior to a point on the frontier curve. The company needs to internalize the concept that continuous improvement has dual purposes: improve operational profit and identify the critical link for process innovation. Such dual concept is important because when the company is operating near the frontier, it would find that incremental improvement is smaller than the incremental effort expanded. Therefore, conventional economic wisdom would argue against continued improvement. However, even marginal benefit is lower than marginal cost, the desire to improve continuously is not so much to increase operational profit in the short term but rather to identify the critical link that would provide an opportunity for process innovation.

In my pre-inked stamp example, the continuous improvements helped me to identify "understanding gel chemistry" as a critical link. When an opportunity surfaced, I immediately saw its importance and quickly conclude the strategic relationship that allowed my company to push the frontier curve locally. Another case example that a student of mine, Pol Narongdej, and I encountered was in a paper making process, where by pushing the production process to operate at the frontier, we discovered certain physical limitations imposed by the dryer's performance [8]. Foster [2] also developed a model for predicting the likelihood for a process innovation by examining an S-shape curve that plots the relationship between the effort put into a process and the results one gets back from that investment. When the company is operating on the frontier point, then its effort in trying to improve the process within the given physical constraint will be leveling off, thus exhibiting an S-shape curve.

A process innovation opportunity can also be found by challenging the "conventional wisdom" on the current industry structure. We start off by asking the question why the industry has the current structure? What are the basic underlying factors that determine the current structure? Does the conventional wisdom believe that these are the key factors for competitive advantage? What are the basic assumptions that give rise to the factors? Focus on the flip side of such assumptions would provide opportunity for process innovation. Take the steel industry as an example. Before the introduction of minimills, steel was like a commodity. There were little productas differentiation and price was the major competitive factor. The reason that the company could not provide product variation was due to the inability of producing a small batch economically using the existing technology. Therefore, mass production was the only viable strategy for a steel company. This led to the conventional wisdom that economy of scale was the major factor for competition [4]. The flip side of this is a new production process that can do small batch production economically, minimills are the answer. In my own case, I went through a detailed reasoning process to identify the opportunity for process innovation: a new production process which is easy to use and can do small batch production quickly and economically.

Another example was Dell Computer's innovation in mail ordering for PCs. When the personal computers were making inroads in the U.S. market, retail distribution specializing in PC products was developed. As the industry grew, PCs became a commodity and heavy competition was on price. The conventional wisdom was to develop the next generation computer faster and make money before competitors copied the same. Dell came in with a different perspective. All the PC producers were facing the same cost structure because all of them had similar technology and relied on the same distribution channel. The conventional wisdom was to compete with better technology. How about competing via channel? The growth of the PC had relied on a specialized channel with technical support. A major reason was that in the initial stage of PC introduction, customers needed technical help in buying a PC. These stores, in turn, required a big discount from the computer manufacturers in order to provide such technical services. The conventional wisdom was that only a retail store with technical sales support could sell a PC. Therefore, when more PC clones came to the market, they competed on the limited PC distribution channel which resulted in reduced profit margin. The conventional wisdom was based on the assumption that the customers knew very little about computers. The flip side of this assumption was that the customers had became very familiar with the computer product. If this assumption was correct, then a different channel might be the answer. Dell found the answer in direct mail ordering.

Potential for process innovation can also be derived from challenging the current practices. An example is provided by the Japanese Just-In-Time (JIT). Inventory control was a long standing problem in the manufacturing of commodity goods. Everyone wants to have zero inventory but this is impossible. With the advances in operations research, optimal inventory control was the hot topic of interest. Sophisticated control algorithm was implemented in the manufacturing plant to reduce inventory cost. Conventional wisdom was that "better inventory control algorithm is the answer to reduce inventory cost". What led to such conventional wisdom? If future demand, supply, and production schedules are known exactly, then we can work out the logistic flow that would eliminate the needs for inventory. Therefore, the need for inventory control algorithm must be due to uncertainties in the OPD chain. So the flip side of the conventional wisdom is to focus on reducing uncertainties in the OPD chain instead of sophisticated control algorithm. The Japanese did precisely this. For example, by establishing distribution channels in different geographical locations, the company could control the total demand by different pricing schemes in different locations. Through design, they create product variations from a core product targeted to different customer groups and control the total demand by different prices for different variations. To reduce supply uncertainty, they established strategic relationships with selected supply vendors and work with them closely to develop an efficient supply infrastructure. Finally, to reduce production uncertainty, they focused on labor relationships, production logistics as well as investments to increase production reliability. With reduced uncertainties, JIT came naturally.

Taguchi Method [9] provides another example. Quality had been viewed traditionally as the deviation of manufactured product from its engineering specifications. With such a perspective, quality control to removing products which are not manufactured up to specification standard. The manufacturing process is designed to produce products efficiently with given specifications and therefore is fixed. Production of substandard products is a statistical property of the given manufacturing process. Under these assumptions, the standard practice for quality control is to inspect products after manufacturing and regard a product as defective if it does not pass a certain threshold level. Taguchi challenged the assumption that the manufacturing process is fixed. Why can we turn our attention to "tune" the manufacturing process such that it minimizes the sensitivity of manufacturing conditions to product output specification. If we can do this every time before making a large production run, then we can almost assure that most produced products are up to specification standard. This lead to the idea of using a small batch as an experiment to obtain information that can be utilized to tune the manufacturing process before making a large production run. This is basically the spirit behind Taguchi Method.

3.2 Creating and Implementing of an Innovative Solution

Once the opportunity for process innovation is identified, the company needs to focus its resource to create an innovative solution to exploit the opportunity. In many cases, when such an opportunity is identified, the innovative solution is rather obvious and conceptually simple. In some cases, the technology and/or infrastructure necessary to implement the solution are not available. In which case, the company is in a "look out" mode which will allow it to react quickly when such missing ingredients become available. This was the case in my pre-ink stamp business. We had gone through more than a year to continuously improve our product and realized that to further push the frontier curve, we needed to understand the chemistry of the gel material. We knew that such knowledge was available but the cost of acquiring it was too high, and therefore there was not much that we could do. However, I was in an alert state. When the chemist called, I could act very quickly to close the strategic arrangement without extensive analysis and evaluations.

In the process of implementing the simple innovative solution, one may find it necessary to change some or all of the OPD chain in order to fully exploit the creative solution. Take minimills as an example, to exploit the minimills solution, the company must retrain its sales force and learn how to do pricing based on customer's needs. To exploit mail ordering of the PC, Dell needed not only to set up the mail ordering OPD chain, it also needed to set up a new infrastructure of phone support to service end user customers. To realize JIT, Japanese companies developed an international sales strategy, changed the product design to allow efficient manufacturing of product variations, implemented an information system that provides fast market feedback, and established a strategy relationship with suppliers. In fact, the more changes required to exploit the innovative solution, the harder it is for competitors to appreciate the "spirit" behind the innovative solution and therefore, less likely that they could copy the innovation. For example, while PC direct mail ordering is easily copied by Dell's competitors, Japanese's JIT is harder to be copied. As a matter of fact, many manufacturers outside Japan had copied the "form" of JIT but missed the "spirit" behind it.

3.3 Expansion Process: The Key to Success

To be able to implement an innovative solution successfully it is necessary but not sufficient for the success of the innovation process. The key to success lies in the ability of the innovator to "scale" up the implementation of the innovative solution quickly before the competitors learn the secret and copy the same. In the middle of 1960, Du Pont developed a new process of making polyester tires which was superior to the nylon tires. Nylon tires were one of Du Pont's major businesses. Instead of scaling up the production and sale of polyester tires, which would hurt its nylon tire sales, Du Pont tried to "control" the introduction of polyester tires so that it could phase out its investments in nylon tire production facilities.

As such, it failed to capture the full economic benefit of such a new process quickly. Celanese, having no nylon tire production facility, took the commanding position by turning out polyester tires quickly and captured over 75% of the U.S. tire market in the late 1960s [2]. The failure of Du Pont to expand the innovation solution provided Celanese an opportunity to establish the strong position quickly.

Another example is given by the McDonald's System, Inc. founded by Ray Kroc in 1955. In the early 1950s, there were numerous franchising fast food chains before McDonald's was founded: among them were Burger King, Dairy Queen, etc. Each of them followed the same formula. The founder created a new process to provide a successful fast food service in a single location. The next step was to franchise this process to other locations. However, each of them stopped growing when a certain size was reached. Similarly, Kroc "discovered" a successful fast hamburger store in San Bernardino, a town in Southern California, called McDonald's, which was run by the McDonald brothers. He licensed from the McDonald brothers the fast food process and created the McDonald's System, Inc. to franchise the process nationwide [10]. Just like any other fast food chain, Kroc started with an innovative fast food process. However, McDonald's System, Inc. differed from other fast food chains in that it enjoyed exponential growth for years. This phenomenal success did not hinge on the fact that Kroc had a much superior fast food process than its competitors, but rather on the fact that Kroc found a creative way to scale up very quickly the opening of new stores nationwide, franchising the same process.

To exploit the advantage of the new process, the company would need to change the way it did business and develop a new infrastructure to support the expansion. However, this implies risk to the organization who is used to doing business in the old way and has invested in an old infrastructure to support it. What makes the company think that the new process would be successful? Would the benefit be greater than the sunk investment in the old infrastructure? The more successful a company is in doing business the old way, the less likely that it will plunge into the new process. Based on conventional financial analysis, a less risky option is to "pace" introduction of a new process to slowly replace the old process. However, in a competitive environment, where every player is searching for creative innovative solutions, an aggressive implementation and expansion of the innovative solution might be less risky than the conventional wisdom of "pacing" introduction of a new process. For example, Du Pont tried to phase out the investment made in manufacturing nylon tires, which gave Celanese an opportunity to dominate the market in polyester tires.

A company can enjoy a short term success if it implements the solution quickly. However, whether the company can fully exploit the potential of such an innovative solution hinges on its capability in developing an infrastructure quickly that supports scaling up of such implementation as well as a strategy that will remove the limitation to continuous growth. To understand how to scale up an innovative solution successfully, one needs to introduce the concept of *Grabber* and *Holder*.

A grabber is a vision of how the innovative solution can provide competitive position. It appeals to one's emotion to adopt the innovative solution. The holder is something that can allow one to actually realize the competitive position as projected by the vision, the grabber. The holder can be a particular product, another process, a specific strategy, or a specific infrastructure. The effect of a grabber is to grab "new" people to try out the innovative solution, whereas the effect of the holder is to retain loyal customers. The feedback dynamic is that if there is a strong holder, then the word of mouth effect will increase the grabber effect dynamically. The result is a self-sustaining perpetual growth. If, however, the holder is weak, then there is no positive feedback in increasing the grabber effect, and the growth is completely determined by the effort in "marketing" the grabber. In many cases, such marketing effort runs into diminishing return and thus limits the growth. If there is no holder, then a "good idea" will never take off and will die. This is described in Figure 3.

When early fast food chains were developed, the franchiser scaled the fast food operation by selling to investors the rights to use its fast food process in certain geographic locations. The franchisee was also required to buy supplies from the franchiser who usually marked up a certain percentage. The grabber is a fast food process that can serve freshly cooked food quickly. This grabbed a lot of investors who believed there would be a latent demand for such a service. The franchiser made its money in selling the rights and marking up in food cost. However, this implied higher investment and operating cost for the franchisee and as a result, the success percentage of the franchisee was not high. While the investor was grabbed by the vision, the real objective of the investors was to make money. The low success percentage implies that the expansion strategy chosen by the franchiser provided a weak holder. The franchiser found it harder to sell additional franchising operations as more units were sold. This builds in a limitation to continuous growth.

Kroc adapted a completely different franchising strategy. He charged very little up front franchising fee and did not mark up food cost to the franchisee. He focused at maintaining a high standard in food quality and cleanliness in all the franchised stores. Because of the cost structure, all McDonald's stores could serve high quality fast food at a low cost. The success percentage of the franchisee was very high. Kroc had developed a strategy that provided a strong holder for owning a McDonald's store. The financial success of the franchisees attracted a growing number of new franchisees, thus creating a positive feedback structure that led to its hyper growth.

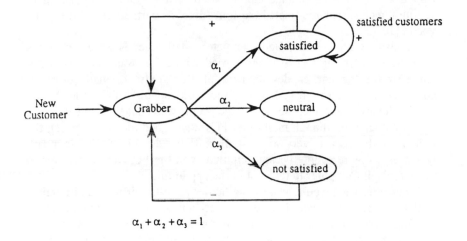

$$\alpha_1 + \alpha_2 + \alpha_3 = 1$$

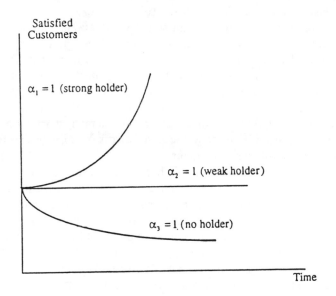

Figure 3. Grabbler-Holder Dynamic.

The interesting question is "where did McDonald's make its money?" McDonald's made its money from the real estate that it rented to the franchisees with a percentage clause based on the store sale. Instead of selling franchisees, Kroc treated the franchisees as partners in fulfilling the vision [10].

Another example is the Dell Computer. Mail order is an old business idea. Mail order for PCs was even tried before but was not successful. However, in the earlier development of PCs, only a small group of customers were knowledgeable enough to "risk" buying a PC through mail order. Moreover, at the earlier stage, the people who could support PC sales were limited. Even though mail order PCs was a good idea (a grabber), the holder for this idea is very weak at the earlier stage of PC development. However, as more people became familiar with the PC, the holder was in place for Dell Computer to the exploit the opportunity.

To identify what would be a strong holder is rather difficult and requires trial and error. However, it is easy to recognize that a strong holder is missing. One basic test is: "if the thing does not take off after some reasonably high effort, then most likely, a strong holder is absent and the system has negative feedback." In general, strategy focusing at passing a good fraction of the benefit derived from innovation to customers would create a positive feedback, whereas, strategy focusing at reaping most of the profit quickly would result in a negative feedback. Also, a complex innovative solution that requires high skill level to execute would tend to create an inherent negative feedback structure, whereas a simple one that can be executed by average skill level would create a positive feedback structure. Very often, the company might try many times before it discovers a strong holder. Once it is found, the company would know it right away because it would take off like a rocket. If the company could not implement an expansion strategy with a strong holder soon enough, it would give its competitor an opportunity to copy the innovative solution and execute an expansion strategy with a strong holder.

Once the company realizes that it has an expansion strategy with a strong holder, then it must focus its resources to build an infrastructure to enhance the implementation of such strategy. Such infrastructure includes corporate vision, organization, culture, information system, reward incentives, etc. The people within the company would develop new skills and acquire new knowledge while implementing the strategy. Eventually, the new competitive edge of the company is provided by the new skills, new knowledge, and new infrastructure developed through the process of implementing and perfecting such an expansion strategy, not the specific innovative solution.

4. INFORMATION TECHNOLOGY: AN ENABLING
TECHNOLOGY TO PROMOTE PROCESS INNOVATION

We can summarize the previous Sections discussions in Table 1. It is noted that to manage the creative activities in process innovation, the whole organization must have a perspective different from operational efficiency: process innovation is a result of human creative activities rather than mathematical optimization. The willingness to challenge the conventional wisdom is the proper perspective. Process innovation is a problem solving activity, but the focus is not on how to solve a problem "optimally" in some mathematical sense but to search for the "right" problem to solve. Once we find out what problem to solve, the innovative solution should be simple in order that it can be scaled up quickly.

From Table 1, we see that the three subprocesses in process innovation involve a lot of searching, analyzing, and problem solving activities. It implies that information technology, which is the basis to support information activities, can be an important enabling technology that can promote and support the process innovation activities. In this section, I like to discuss how information technology should be managed properly to achieve this goal. Before the discussion, I like to point out that it is people, not information technology, who create the process innovation activities. People's perspective is the key driver behind such innovative activities, and therefore it is important to first understand the role of information technology in influencing people's perception in thinking about problems.

We can broadly classify information technology into two classes. The first class is focused at organizing, retrieving, communicating, and displaying "raw" data. The second class is focusing at interpreting, analyzing, and manipulating raw data to provide useful information for decision making. Data base, networking, telecommunication, multimedia, are technologies in the first class. Statistical analysis, imagery analysis, optimization algorithms, expert systems, neuronets, are technologies in the second class. We shall refer to the first class as Data Technology, and the second class Problem Solving Technology. Data Technology tends to influence people to think in terms of how such technology can improve the activities that we are already engaging in. Examples are electronic order processing, desk top publishing, teleconferencing, home banking, computer base training, and many others.

Current Problem Solving Technology is primarily focusing at solving problems while assuming that the OPD chain structure remains the same. For example, statistical analysis is to do the "best fitting" of parameters to observed data collected under the existing structure. Sophisticated scheduling algorithms are developed to optimize efficiency for existing logistic structures.

Table 1. Summary of three subprocesses in process innovation.

Subprocesses	Perspective	Activities
Finding Potential for Process Innovation	Challenge conventional wisdom in current cost structure, market structure and practice	Continuously improve operation to identify limitation, market structure analysis, modeling
Creating and Implementing Innovative Solution	Solution is clear once the "spirit" behind innovation is identified; implementation requires changes in existing OPD chain	Problem Solving: analytical method or case reasoning
Scaling up Innovative Solution	Expanding the innovative solution quickly is less risky than "pacing" its introduction; Grabber-Holder dynamic	Search for a strong holder; Infrastructure development

Table 2. Use of Information Technology.

Information Technology	Conventional Use	Promote Process Innovation
Data Technology	Improve or Replace Current Activities	Understand Industry Structure
Optimization Technology	Find Optimal Solution	Explore Limitation to Further Improvements
Modeling Technology	Forecasting and Planning	Derive Insights in Industry Dynamics and Players' Behavior
Expert System Technology	Put Expert in a "box"	Increase Knowledge level of people
Case Base Reasoning Technology	Support Tactical Decisions	Support Human's Activities in Problem Solving
Client-Server Technology	Support Distributed Operational Activities	Support Distribution of Knowledge and Insights
User Interface Technology	Interactive data retrieval	Interactive Problem Solving

Complex production planning systems are developed to meet projected future demand, assuming a certain fluctuation and a given value of chain activities. Fast algorithms are developed to solve very large scale mathematical optimization problems, and expert systems are developed to put experts in a "box".

The current use of information technology tends to influence people to have the perspective that the way to solve existing problems is to improve efficiency under the existing structure. The push is towards automating the activities that we are doing now. However, from Table 1, we see that this kind of thinking might hinder our creativity in process innovation because creativity starts from requiring us to challenge our existing structure. The more comfortable we are with the efficiency approach to problem solving, the more risky it would feel adopting a new structure. Under such a situation, even though an innovative solution is found, our chance of implementing it successfully would be small. From the management point of view, we need to redirect the use of information technology such that it will not only help us to do our current activities more efficiently; it will also influence people to think in a broader perspective, to challenge the existing structure and to look for ways to create new structure. The push should be less towards automating current activities but more towards helping people to be more creative.

4.1 Finding the Problem Instead of the Solution

Current use of information technology is primarily to support finding data or/and solutions to a problem. However, in the search for innovation potential, we are more concerned with finding "the problem" that can provide the opportunity. As we discussed in Section 3, understanding the current structure, the industry structure, and practices is the starting point of finding the problem that can offer innovation opportunity. This sounds very simple but, in fact, is a fatal mistake made by many corporations. Many of the historical cases where the market leader was overtaken by new entrants were due to the fact that while the leader was still operating under the existing structure, the new entrants were riding on the benefit of an innovation which changed the existing structure. Examples are U.S. steel industries in the 1970s and U.S. consumer electronics industries. The first issue is how can we direct the use of information technology to support finding the problem instead of data or solution.

As discussed in Section 3.1, finding the problem that leads to the limitation in the cost structure starts with pushing the operation to a point close to the frontier curve. From an optimization point of view, this means that we care more about discovering the binding constraints of a feasible set rather than finding the optimal solution quickly and accurately. Thus, we see that optimization algorithm, even though commonly viewed as a technology, aims at finding the optimal solution quickly and accurately, can also be exploited to provide hints in discovering the limitation that causes the existing cost structure. However, the algorithm and the way the user should interact with it must be changed to emphasize discovery of binding constraints rather than finding the optimal solution. For example, one can use duality property of an optimization algorithm [11] to develop a graphical presentation of how the ratio of marginal benefit vs. marginal cost can be influenced by the adjustment of certain processes or physical parameters. This can greatly enhance our capability in visualizing which parameter is more effective in pushing the frontier curve. With the advances in powerful PCs, we can further develop an interactive program that can allow one to explore and find the critical link that can push the frontier curve in the right direction.

Marketing research, modeling technology, and optimization algorithms have been used extensively to support sales forecasting and production planning. The trend has been towards building more complex econometric and/or dynamic models in order to be comprehensive enough to simulate all the possible situations. As a consequence, these models become less "transparent" and even the model developer cannot fully understand the structural characteristics of the situation being modeled. However, to look for process innovation potential, we are more interested in understanding the industry structure and what are the basic assumptions that cause this structure. Therefore, the push should be to use marketing research, modeling technology, and optimization algorithms to help us understand the industry structure and the assumptions that lead to such structure rather than forecasting and production planning. For example, we can develop a software program that organizes marketing research data and an interactive program that supports an analyst to retrieve and perform analysis on the data to verify certain hypothesis for the current market structure. For example, is it the cost structure of different players in the industry, the income distribution, or other variables that leads to the current industry structure? We can also use a system modeling package to develop simple dynamic and optimization models to provide insights about dynamic interactions of current players and discover the basic assumptions that determine the stability of the current market structure.

4.2 Upgrading Knowledge within the Organization

In the search for limitation in pushing the frontier curves, people's knowledge on the existing operation plays an important role. In several experiences that I had, bringing up the knowledge and skill level of the people within the organization was the first step to allow management to motivate people to push the operating process to its limit. The risk profile of a person can be changed by education and training.

In a paper mill plant in Thailand, Narongdej and I developed a system, incorporating expert system and multimedia technologies, that would improve the knowledge level of all the plant workers. The system provided confidence to the unskilled operators to engage in pushing the process to operate on the frontier curve. One emphasis in the design of the system was the ease to upgrade knowledge as more experiences were accumulated. The other emphasis was the user interaction in getting knowledge and recommendation from the system. Since most of the workers had no prior experience with computers, we used multimedia technology extensively to develop displays with pictures, diagrams, and tables. The systems developed were very different from the conventional expert system approach where the focus is to "replicate" the expert's knowledge [12]. That system became an integral part of the continuous knowledge accumulated in the plant. The result was that the plant people discovered the limitations of the existing production process much faster [8]. The design approach was based on a general approach discussed in an earlier publication [13]. In that paper, I discussed, in general, how expert system technology should be modified so that it can be used to continuously improve the knowledge of the operators in the process of pushing the operating performance of a manufacturing environment.

4.3 Building up a Case Base for Success and Failure of Process Innovation

Case base reasoning [14] is a new technology used to support operational decisions and planning. The technology was developed based on the assumption that many people solve new problems by associating with how they have solved similar problems before. A case-based reasoning software organizes historical cases on how certain problems were solved and, via query, allows one to retrieve these cases. When a new situation appears, one can input appropriate questions that will retrieve all the relevant cases and how they were solved. With a library of cases based on experts problem solving experience, the technology can support people with less experience to deal with unfamiliar operations. The technology also focuses on how to modify old cases, combine cases which are similar, and add new cases which are different in nature from the old cases.

We can apply this technology to develop a solution concept for an innovation opportunity that can further be analyzed and refined. To proceed in this direction, we first build a library of cases on the success and failure of process innovations. This library should cover not only those in different industries but also those within the company. In addition to the full case studies, we should also extract the ideas behind each of these cases: e.g., the basic idea of JIT is "to reduce uncertainties in the OPD process is a better way than solving an inventory problem", the basic idea in McDonald's success is "to share the benefit of process innovation with those who will spread the innovation will create a strong holder", and the basic idea in Du Pont's introduction of polyester tires is "to pace the introduction of an innovative solution can be very risky". Using advanced text storage and retrieval technology, we can document and organize these historical cases and their associated ideas so that one can retrieve the relevant case easily through a "concept" or "topic" based text retrieval system. This system can support problem solving activities in finding an innovative solution and a strong holder that can support the expansion of the solution.

4.4 Flexible Infrastructure to Support Change

If the company anticipates high frequency in process innovation, then it must accept structural change as a norm instead of a rare event. This means that the information system installed must be flexible enough to support frequent structural change. Data, knowledge, cases, and problem solving capability required in the new structure may be different from those in the old structure. Therefore, the information system installed must allow us to modify and reorganize data as well as knowledge very quickly at low cost in response to change in structure. It must also allow us to develop or modify problem solving tools quickly at low cost in response to change. Cost and efficiency in supporting the current structure are not the only attributes of concern; the manager needs to incorporate the attributes of maintainability, compatibility, transportability, and upgradability in the decision to install an information system. In other words, we may need to pay some premium to develop information with greater flexibility. An example of such a system is based on the recent development in client-server and object oriented technologies which can provide flexibility in general object organization, management, and distribution.

In Table 2, we are summarizing how the existing advanced information technology should be redirected to promote process innovation. We see that many technologies, which were developed to support efficient operation of current information activities, can be modified to enhance the subprocesses in process innovation.

5. CONCLUSION

Process innovation can be decomposed into three subprocesses: a search process, a solution process, and a change management process. In the search subprocess, we identify the opportunity for process innovation. In the solution process, we develop an innovative solution that can exploit the opportunity. In the process of scaling up the innovative solution, the company needs to change its way of doing business. A new infrastructure needs to be developed to support the expansion process. Successful implementation of these changes is the key to successful process innovation. Process innovation is very information intensive and therefore proper management of information technology plays a very important role in supporting process innovation activities. Management needs to understand how information technology can influence people's perspective in thinking about problems. It is pointed out that the current trend in the use of information technology may hinder process innovation. Finally, we recommend how information technology could be redirected to promote process innovation.

REFERENCES

1. **Diebold J.**, The Innovators: The Discoveries, Inventions, and Breakthroughs of Our Time, Truman Tally Books/Plume, New York 1991

2. **Foster N.R.**, Innovation: The Attack's Advantage, Summit Book, New York, New York, 1986

3. **Reid T.R.**, The Chip: The Microelectronics Revolution and the Men Who Made It, New York: Simon & Schuster, 1985

4. **Drucker P.F.**, Innovation and Entrepreneurship, Harper & Row, Publishers, Inc., 1985

5. **Hayes R.H.** and **Wheelwright S.C.**, Restoring Our Competitive Edge; Competing Through Manufacturing, John Wiley & Sons, 1984

6. **Tse E.**, **Zhu M.**, and **Juang S.Y.**, " SIMMEMS: An Intelligent Support System for Manufacturing Enterprise", Proceedings of 2nd International Conference on Computer Integrated Manufacturing, May 21-23, 1990.

7. **Simon H.A.**, The New Science of Management Decision, Englewood Cliffs, NJ, Prentice-Hall, 1977

8. **Narongdej P.**, Interactive Man-Machine Learning in Dynamic Manufacturing Environment, Ph. D. Thesis, Department of Engineering-Economic Systems, Stanford University, Stanford, California, 1993.

9. **Taguchi G.** and **Phadke M.S.** "Quality Engineering Through Design Optimization" conference Record, vol. 3, IEEE Globecon, November 1984, pp1106-13

10. **Love J.F.**, McDonald's: Behind the Arches, Bantam Books, New York, 1986

11. **Luenberger D.G.**, Introduction to Linear and Nonlinear Programming, Addison-Wesley Publishing company, 1973

12. **Alty J.L.** and **Coombs N.J.**, Expert Systems: Concept and Examples, NCC, Manchester, 1984

13. **Tse E.**, "An Intelligent Support System for Continuous Improvement in a Dynamic Manufacturing Environment", Computer-Integrated Manuf. Systems, 1991, pp 229-238
14. **Shank R.C.**, Inside Case-Based Explanation, Hillsdale, NJ, L. Erlbaum, 1994

Chapter 12

MANAGING INNOVATION IN CIM ENVIRONMENTS

Sam Bansal
GINTIC Institute of Manufacturing Technology
NTU, Nanyang Avenue, Singapore 2263.

1. INTRODUCTION

The economics justification process has long been identified as the biggest hurdle to the adoption of advanced automated manufacturing technology [8]. In recent years, the literature has been inundated with a large number of methodologies and evaluation techniques that look promising for the economic justification process for advanced automation technology [1 - 7]. Excellent books have been written on the topics of justification [14], innovation [13], change, and change management [9,15,19,21,22].

Managing is a difficult process no matter where. For "supportive innovation" efforts, however, it assumes a far greater seriousness. This is because for such areas the driving influences are numerous and training and tools not enough. It has been seen and heard, there was a time in mid eighties, that CIM technology was going to save manufacturing. Then came TQM, DFM, QFD, CE, and so on. Lately, from every corner, comes the rallying cry to Reduce Time to Market, Improve Quality, Increase Yields. Empowerment, C++, OODL, and others are such disciplines that proponents of them never tire to mention their virtues. Amidst so may variables that may be driving the practice of innovative technologies, that management of them becomes a very daunting task. Trained personnel, managers and implementors, are not readily available. Nor are the tools that can make the life a bit easier. For the longest period, educational institutions have not done much effectively to ease the pain.

The question then, one may pose, is why is it so? Having reflected for some time on this issue, one may come to the conclusion, that management of innovation in CIM environments is tougher because of certain structural reasons. Looking at Figure 1, one sees that customer based demand influences the product design, which in turn impacts the manufacturing process and it in turn drives the "supportive innovations". One such is CIM.

CIM technology, and its practitioners, managers, and results, if successful, will influence in turn, the upper stream domains, i.e., manufacturing, design, and maybe even customer generated demand.

It has been demonstrated that innovative management in CIM environments is different because it is structurally placed at the tail end of the value chain. Then there are too many upstream domains and their needs that drive this subordinate process. Lack of trained managers and practitioners without the aid of scientific tools, who have been practicing these technologies as an art, have had a sporadic success rate. Technologies have been around, but the science of their integration has not been.

In the rest of this paper, the justification and innovation management processes in CIM environments will be explored.

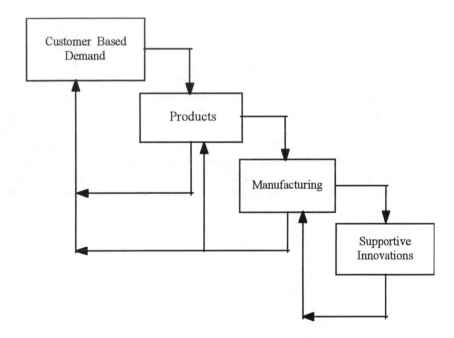

Figure 1. Linkages with Feedback Loops.

2. BACKGROUND OF TWO CASES

The present paper is written based on this author's personal experience of having done two mega projects [10,18] involved with the above domains. Other pertinent papers dealing with these issues have appeared earlier [11,12,16,17,20,23]. One of these two projects was a Plant Wide Automation [18] of a continuous and batch process plant of Polaroid Corporation. The other project was the Enterprise wide CIM [10] at Alcoa Electronic Packaging. Both were 5 year projects requiring justification of upwards of $20 million. This paper is based on the actual life experience in justifying and building innovation.

In both instances, management was partially disposed to supporting innovation. But, true to the form of management they were looking for answers to "prove it to me" syndrome. They were, in both cases, looking for a sure bet, a winning horse in the form of a champion. As was proven later on, if they did not terminate the discussion in the first meeting, there was some hope. But it had to be backed by facts and hard work. Both environments were hard nosed if you, the champion, were not sufficiently thick skinned, you would last the consistent and continual grueling.

Were these environments any different than one would normally come across for major projects? No they were not. However, based on academic theories, one would not be able to justify the megaprojects there. What was required in these cases was the persistent articulation, as will be described in this paper later on.

These lessons were confirmed even later on as the practice went on to be carried in the Asian region.

Specific company details are provided elsewhere [10,18]. Point being, what management was looking for not solutions to some equations but were the factors described in this paper in the following pages. Nevertheless, the systems realization of the champion's dreams are given in Figures 2 and 3.

3. JUSTIFICATION PROCESS

If the senior management of the company is unimaginative, reactionary, complacent, or non-risk-taker type, they would be unresponsive to innovation. Unless the competition or the forces of the market change them, they will not listen to anyone. In another situation, unless the senior management is predisposed to implement the subject innovative technology, it has got to be justified. And perhaps because of the "value chain concept", or whatever else, the justification process is very difficult if not impossible. After a long time in doing these justifications, few thoughts have gelled, from the perspective of the senior management and also from the perspective of the implementor. They are given below.

3.1. Management Perspective

Unless the management itself is subject matter expert, the issues that they are concerned with are: is due diligence done? Is it truly justifiable? Who is competing for funds? Is there a champion? Is there a sponsor? What is the risk to the sponsor, to the company? Is it better than some product development program? What is the level of comfort of the sponsor with champions and the technology? Can I trust these people? What does it do for my image, plan, or future? Does it conflict with any of the goals of my boss/company? Does it help, any of the product, process, material, or yield improvement programs? Does it help any of my critical problems with people, material, space, and profitability, i.e., does it lower my direct or indirect costs? If so, can I live while they develop and complete the implementation of the promised technology? And will it put me in a bind after implementation is finished?

Is Due Diligence Done?

A question usually asked, is due diligence done? Are all the staff groups in agreement and supportive?

When the senior management is approached for review and support, sometimes they may begin questioning the proposer. Not that they know the subject technology better than the proposer, but it is some to learn and some to judge how comfortable the proposer is in answering the questions. At the end of this session, if the proposer did well, most likely, he will be told to seek the support of the staff groups. This command can take many forms. However, in all cases meaning is the same. The staff groups, in normal circumstances are, accounting, industrial engineering, all relevant operating and support departments. Once again, the senior management is looking for a consensus opinion that the plan is well thought out. While it may lack precision, all angles have been considered.

Before one embarks upon the review with the staff groups, planning and design has to be done. This entails collection of functional requirements, translation of the same into systems requirements, and responsive architectures. These can be documented and sent to generate the RFPs, if appropriate. System requirements and the architectures form the basis of the cost and schedule estimates. Training and technology, transfer plans, and procedures are thought out and put in place. While doing the requirements definition, benefit compilation can also be done. This work can be compiled in a general design document and be used for presentation to the senior management.

It would be prudent, time permitting, to review the proposal with the staff groups. They are either going to express their support, disagree, or argue, to no end. If they point out a problem and the proposer agrees, care must be taken before proceeding further. If they argue, politely back out of it. If however, disagreement is genuine and the proposer does not agree with it, he should note it and still proceed. It may have to be resolved with the senior management during the review process. He has, at least, even odds to win his case.

Having reviewed with all the staff groups, due diligence can be considered done and he is ready to go to the senior management.

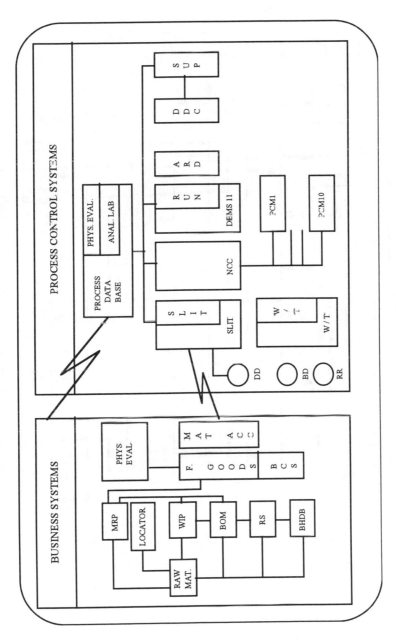

Figure 2. New Bedford Computer Project.

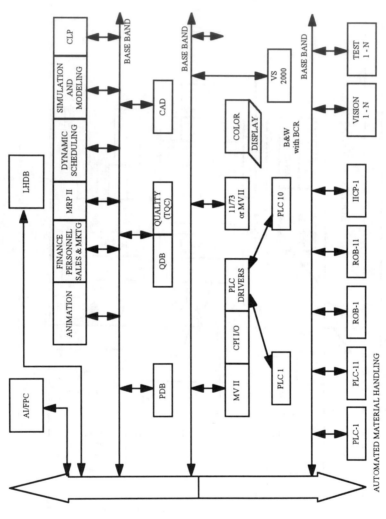

Figure 3. ALCOA Electronic Packaging's CIM architecture.

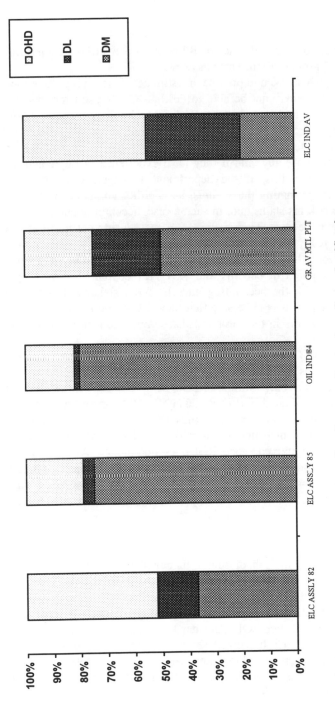

Figure 4. Cost Structure Comparison Group Justification.

Is It Justifiable?

It was mentioned during the due diligence process to start compiling the benefit. Staff groups are the same as above.

However, when asked to provide an estimate of the savings, due to a new technology, they may not be able to do that. One reason for this is: they genuinely do not understand how the subject technology's incorporation will cause savings. Proposer may not know it himself, or cannot articulate it. In such a case, outside help should be sought from the consultants or the friends, if there are any, and if they have experience in implementing the same. Even after this, if the department manager, does not want to participate in estimating the savings, he is either not capable of it or does not want to risk his status quo. In such a case, accounting and/or industrial engineers may be tried. They may help or may decide not to. In any event, proposer is ready for the senior management review.

Having gone through all relevant departments to be affected, a summary should be prepared of benefits and ROI [12] computed. Next step is to review these with the accounting and the industrial engineering groups. Main purpose of this is to be sure that they will support and that there are no systematic errors in the proposal that may cause it to be shot down in the final review.

Utmost caution should be exercised against the temptation to pick some savings estimates from some published article or book. The main point of this exercise is not only estimating the saving, but even more important is the commitment of the department manager, that he knows what he will get and how will he use it to get the savings, the two of them are estimating.

An ROI of 40% or better is generally considered good. However, if the company is cash starved, good ROI does not matter. Also, if alternative investment are producing much better ROIs, then the subject plan may not be attractive. In such cases, strong considerations have to be made of intangibles, such as the cost of lost opportunity, lost market share, etc.

Who Is Competing For Funds?

It is of course the job of senior management to weigh all the facts, factors, and constituencies. Proposer may not know who else is competing for the funds and what is the strategic importance of investments in them, opposed to funding his proposal. The competing requirement may be producing even a better ROI. Consider for instance Figure 4. In this, five cases are shown. Each bar graph shows direct material, direct labor,. and overhead costs. These added together make up the total cost to manufacture the goods. Case 1 is of an electrical assembly plant costs in 1982. Case 2, shows the same plant after CIMing it completely. Case 3, is that of oil industry averages, representing the best cost structure, among the five compared. Case 4, is of a metals company average, whereas Case 5 is of electrical industry averages. In case 4, direct labor is 25% of the costs. The high labor component is not due to high number of people employed, but is due to high wages and benefits paid to them. In this plant, cost to further automate to reduce labor would be very unproductive. Also of the 25% overhead costs, 18% were recurring costs, due to allocation of depreciation, etc. and as such are not easily impeccable by automation. Consider under this scenario, if another plant were to be built. If this plant is built, elsewhere, where cost of labor is quarter or less and where due to corporate rules 18% recurring cost is not allocated to the plant. Simple math leads savings of 37 %. These are very easily realizable. And especially, if a new plant is to be built. No wonder, so many companies have gone to the lower labor cost areas, instead of improving the existing operations with the automation.

If such is going to be the case, in a retrofit situation it is very hard to refute the alternative. In a green field case however, the argument would be on the basis of cost avoidance and building even a healthier enterprise from cost point of view. Consider asking the staff groups these questions, while reviewing with them during the due diligence process. Most likely they will tell you about "who/what else". This may prepare one to argue with senior management. They may or may may not allow such debate or even be disposed to listening to the proposer's side. But in doing so, nothing has been lost. Actually the proposer has gained an additional knowledge of the company's business and its plans.

Is There a Champion?

Most big projects have a champion. In the context of this paper, the words proposer and champion have been used interchangeably. This is the person, who is fully and personally committed to seeing the project through. senior management likes to see one such a person, for every major project. If there is going to be a major innovation project undertaken, a champion has to come forward. Head of a committee is not a good enough champion. A true zealot would be one. Contrary to popular opinion, even if this person is from the subject technology area, as a manager or a subject matter expert, with enthusiasm and commitment, management would accept him. In our experience, it is seldom that a major innovation project was approved without a champion supporting it.

What are then the credentials of such an integral member of the team? Needless to say, this man must have unquestioned credibility with senior management. He should be well respected within his peer group for his knowledge or vision. He should, as well, be a can do kind of a change agent. Technically, if this person was capable of doing all the due diligence, justification, compilation, and analysis, the project would benefit immensely. Above all, this person has to know what is the company's business and the division's manufacturing strategy. He should be in a communication loop, that he will be well informed of the changes to the strategy taking place. Of course, he has got to be a good project manager.

Is There A Sponsor?

Champion alone will not do. There has got to be a sponsor, who is either a mentor to the champion or is personally committed to the subject innovation. Best of all, if he is the head of the manufacturing division. He does not know how to implement the subject innovation, but feels the need very vividly in his mind. He is well aware of the virtues of the innovation and knows that without it his operation would face tough times. He is the person who will wield enough influence to set up the time to meet his superiors, for the champion, and the project team. He is also the person, who will provide support when the project will run into tough yet resolvable times. In short, he is the top management person without whose support champion is powerless. He is the person who walks the project through the fiscal processes of the company. He is also the person who can arbitrate priorities for his division for capital spending and expense budgets.

What Is The Risk To The Company, To The Sponsor?

This is a very subjective issue. If asked, every sponsor may deny any practice of it. However, the fact is, every human being practices it. They may not want to accept it though. Objectively speaking, what this entails simply is if there is any determinable risk that the subject people cannot deal with. Either due to financial or personal loss of some type. Uncertainty of benefits or project cost control may bring bad name. Supporting very unpopular company projects would be another category of risky ventures. Every sponsor evaluates and walks with a certain amount of trepidation in areas personally unfamiliar to him.

Champion's task obviously is to allay most of such concerns, in a very unobtrusive, courteous, staffly, and objective manner. Personal relationships or outside education and training done with finesse can alleviate some of these problems. If there is a genuine risk, be candid and up front about it. They would appreciate it. Do not hide any such facts. But then, offer and be prepared to deal with the uncertainties and the risks.

Is It Better Than Product Development Programs?

If the project is compared against some product development programs, it is very tough going. It is not fair for the project either, especially because of innovation projects placement in the value chain. However, most good companies will have the sense not to compare these against the product development programs, but look at them in the manufacturing upgrade, facilities upgrade, equipment modernization, productivity improvement, or some such category. If the staff person, responsible to review, does not categorize appropriately, reasoning should be attempted with him. This is another place where champion's relationship and credibility can pay off.

Level Of Comfort Of The Sponsor With The Technology, Champion, Team

If the sponsor is the least bit uncomfortable with any of the above, the discomfort has to be changed. If he is uncomfortable with all three, champion has to alleviate it. Usually, they will be mostly uncomfortable with the technology. Though in their heart of hearts they know that they have to adopt the technology. Here again champion can pay his dues. Teach the sponsors, if they provide an opportunity and convert their discomfort into a comfortable experience.

Can I Trust The Leader, The Team?

This situation is much more aggravated, if the champion is new. Mostly, he will not be given the chance. Yet, if he is, which happens frequently due to circumstances, they do not trust him They do not believe what he says. They do not know the technology themselves, and are not ready to accept what is being said by the champion, and especially if it is in conflict with one of their old time employees. Situation is hard, but a good champion can take care of it. Tools that he has at his disposal are, candor, sincerity, knowledge, desire to help, and above all to work the process with patience and perseverance.

How Does It Link With Our Strategy?

Questions on the minds of senior management are: What does it do for my image, plan, or future? Does it conflict with any of my bosses goals? Does it help any of our strategic products, processes, materials or yield improvement plans? Does it help any of our critical problems with people, materials, machines, spaces or profitability by impacting direct or indirect cost? Essentially the inquisition going on, whether debated or not is, what does it conflict with, if any? And, what does it support, if any? The linkage that they are trying to establish is with the company business or division manufacturing strategies.

The champion challenge is obviously to relate as much of the innovation plan to the "strategy" as possible. This involves knowing, understanding, articulating, and translating the company strategic plans and then relating all or parts of the innovation plans to it. Innovation plans functionality has to be responsive and supportive to the company/division strategies. In addition to functionality, the plans implementation and performance time has also got to be consistent with that of the company/division's strategies. Once successfully done, it must be kept in synch with the company's strategies at all times.

If the strategic linkage is well established, it should be highlighted at all available opportunities, to win friends and support. If communicated and articulated well, it would satisfy most of the mind rattling queries in the minds of management, and they would be highly supportive.

Can The Operation Deal With the Innovation During Implementation and After?

The last of the hurdles management has to cross, is to get a comfort level that the subject innovation will not impact the operation harmfully, during and after implementation. Management that is not the subject matter expert, is dealing with a new set of implementors, is highly suspicious, on these grounds. These are very natural human reactions to the unknown. The best the team or the champion can do, is to provide a line by line analysis of the worries management has expressed. Not paying attention to them, or ignoring them, will be counter purpose. Explanation should be given in the analysis, how a worrisome scenario will be satisfactorily handled.

This done well, and done with conviction, will win support from most skeptics.

A worried plant manager cannot articulate his fears. Yet what he is saying, sometimes is, just assure me how my operators will use the systems? And how will I use it? Most of these can best be taken care, if the team would simply develop and present the Current and Future Operations models. However, both the management and the team may not be able to articulate what each other is saying. The gulf keeps on increasing instead of going the other way.

Technologists Sensitivities To Senior Management Needs

Some basic points the team and the leader has to always keep in mind as they deal with the manufacturing and company management are:

These guys are under the gun for relentless reduction in time to market. Quality revolution has added another quantum or two of troubles. Then there is higher yields and reduced cost of doing business. If they cannot achieve all of these, their job would be done for them. They would be divested, sold, or shut down. In any event, there are dangers to the jobs. Downsizing and restructuring is a constant happening, now practically in every country.

In this scenario, if the business and manufacturing management becomes hard-nosed and less patient, it should be understandable. They want to upgrade, modernize, and streamline their facilities, equipment, and plants. Innovation can help. They are not the subject matter experts. They rely on staff groups, like accounting, industrial engineering, and relevant operations groups, to do staff work. These groups, over the years, have developed language and matrices, that are commonly understood. Then there is a common language for the operational problems, such as yield, ROI, capacity, quality, control, time to market, etc. Hence, subject matter experts have to translate their message to language and criteria that management understands. They are not the experts in these fields. So the experts have to do the transcending. This sensitivity is required in all phases of the project, as one deals with business/manufacturing management.

3.2. Technologists Perspective

A young technologist without much experience in dealing with management could get a serious jolt if subjected to the justification process. Seniors not only have to shield and support, but have to train and educate the junior member of the team. However, various emotions we have heard on these excercises have been: justification is a subjective process. Nothing could ever get justified, filtering through the various groups' objections, as laid down before. It is an excercise in futility. I do not know the company strategy, worse, they themselves do not know it. Still worse, it changes too often. Our management is complacent, covering themselves, he is reactionary, and so on so forth.

The fact is, it may all be correct. But then it may not. It is believed that if one follows the justification process and it works, he will find either success, or a real explanation, if management does not accept his plan, the reason behind it.

Undoubtedly, due to placement in the "value chain", justification of innovative technologies is harder. But it is not impossible. It can be done and has been done, provided the process has been thoroughly followed.

It should be realized however that because the proposer/champion wants to do CIM, there is no reason why management should want to do it. While it is a tremendous enabler of the company goals, it is the job of the champion to articulate why and how it is so. CIM strategy should be aligned with the company and business's manufacturing strategy, both for functionality and time frame. Innovation strategy is subordinate and responsive to the company, business, and manufacturing strategy.

4. IMPLEMENTATION

Once again, what is going to be mentioned here, are the items that are believed to be unique to the management of innovative technologies. Hence, the classics such as, project management, etc. will not be mentioned here.

4.1. Planning

As an innovation manager, one is expected not only to plan but plan well. These plans should include regular schedule plans and in addition should have contingency and back up plans. In short, an escape route should be planned for every contingency that can take place. How often should senior management be apprised depends upon the local situation. But the more they are told, the more they worry. And sometimes they can come to wrong conclusions about the project.

4.2. Preventative Management

Innovative projects' management doesn't give enough luxury to go wrong, on the delivery of results or on time. Good innovative managers constantly focus on managing catastrophes before they happen. One way to manage with this spirit is to have contingency and back up planning for every identifiable situation. Second and third, is to manage against schedule and cost overruns. Both of these require precise management of the "change of scope". This creeping paralysis can be caused by the team members or even the customers. However, the key here is to manage, before it has harmed the project.

4.3. Teaming

In straight department management one could get away by the boss/subordinate relationships. However, in these management situations, one has to often deal with a matrix organization. This organization, by its very nature of formation is multidisciplinary, multidivisional, and hence a multiagenda team. Boss ordering to the subordinate will not get the task done.

Hence, how to form, how to run, and how to manage are the usual issues.

The leader or the champion should do his due diligence. Make presentation of it to the candidates interested in collaborating in the effort. Receive permission from the top management to hold a recruitment meeting if they support it. Recruit the needed talent from these meetings. More than one meeting may have to be organized. General meetings should be followed by one to one meetings. The candidate and the leader should independently decide if they are going top work together. Divisional management permission was secured up front, but they have to be informed again and should tell the candidate, when can he join the new effort.

It is a risky proposition for the candidates. If a move is required it is riskier. Initially, invite them to work with you. Salary administration should remain with the old supervisor. It should not change until all are in agreement. Agreeing to join the team is putting a lot of trust in the leader and the articulation of the plans that have been laid in front of them.

The leader's attitude has to be supportive and sensitive to build an integrated team with one agenda and that is, complete implementation with success.

4.4. Communication

Communication once is not enough. It is periodic communication. It should be repeated with some frequency. People attending should include, the team, the customers, the management, and also the vendors. If the rules and logistics do not permit, multiple sessions of the same should be held. Content may change from preliminary plans, to current operations model, to future operations model review. Later on, technology details followed by implementation schedule to technology transfer plans may become the topics to cover. This is not meant to supplant the usual design and schedule review meetings.

4.5. Motivation

A motivated team will climb mountains and create wonders. A good leader and a good champion can fire up his teams to believe it may be the best in the entire enterprise. However, this requires sincerity, sensitivity, and support. The team knows the leader will support them. The leader knows the team will follow his advice and edict, if seldom required. The leader has laid his own job on the line to protect his troops and the troops know it. This has created environments of loyalty, camaraderie, and top workmanship.

Other factors, for keeping morale high are good pay, good benefits, good promotion paths. But for innovative technology workers, sometimes called "gold collar" workers, well laid and well understood innovative plans with also well understood individual roles and good working environments, perhaps are greater requirements.

4.6. Customer Satisfaction

Customers are the reason for any organizations existence. An unsatisfied customer can harm a lot more than a reasonably satisfied customer. But without managing properly, they too can create chaos. One way to get acceptance is by prototype demonstrations, rather than cryptic written documents. Let the prototype demo be the agreement time, with the changes noted and agreed to. Second time they can be difficult, is at the hand over time. This usually means more time and more training. It is not a huge problem. Customers should be accommodated every way they can.

4.7. Plan Totally But Implement Incrementally

A common wisdom is to do it incrementally. There was a time when it would have made sense. That was the time of the giant monolithic computer-based innovative solutions. But now the paradigms of systems sizes have changed. Incremental implementation makes some sense, from a logistics point of view. Nonetheless, to proclaim it as the only way of implementation does not sound correct.

Consider two cases. One is greenfielding and the other is retrofitting. If one is to greenfield, incremental implementation, it would create hurdles that one has to contend with throughout the life of the project. If on the other hand, it is a retrofit situation, then the difficulty will be similar.

If the justification is sound, due diligence is well done and the proposal is well sold to the management, they will not have the patience to wait. Nor should the implementing team. Reduced time to market and demand for the reduced times to implement, practically necessitates, that the whole integration proceed as fast as possible. However, if the team and the leader is on a learning curve, incremental implementation makes sense.

However, what would be the best of all the worlds, is to "plan totally" and implement incrementally, within the tactical constraints of the available funding, resources and the test times.

4.8. Time to Implement

In early eighties, many systems groups had come to a conclusion that management should invest for the long term. They translated that demand from the "gurus", to the CIM systems as well. Quantification, of that was equated to ten years. In practice, management never bought that as the implementation time frame. Two years in those days was acceptable. Now, six months is a magic number. From a team's perspective, on a learning curve albeit fast, such time frames are unacceptable and impractical. Their one answer may be, not possible.

Consider however, the fact that the management itself is under relentless pressure to reduce time to market. Down size, divest, globalize the success, introduce new products, etc. are the realities shaping their actions and reactions. This in turn, is forcing them to have a manufacturing strategy that is very dynamic in time and whose periodicity is short. Under these conditions, to expect that the management would buy a long time frame to complete an innovation job, is next to impossible.

Two solutions are plausible. One is to search and find a highly experienced team leader and give him carte blanche to run as fast as possible. Second, is to implement concurrently. The graphical representation of various stages of implementation is given in Figure 5.

The concurrent idea is to implement as many of these stages as possible, in parallel. Overlapped implementation, should be accepted as the worst alternative. Whereas today's realities will almost never allow the sequential scenario, due to the long time frames that it requires. This is a topic of considerable research interest. The right tools, once developed and put in place would make their impact felt all around.

4.9. Integrated Methodology

Research in the tools that was mentioned above, is concerned with the "development and implementation of integrated methodologies". This is shown in Figure 6.

The idea attempted in the diagram is, a tool that can take the various inputs and can produce from business strategy to a CIM planning and implementation specification, including the required documentation at the intermediate stages.

Such a tool would not only shrink times to implement, it would produce documentation as required at every intermediate stage. It is realized, that this kind of tool is not available currently, but that it is presently under investigation.

5. STRATEGIES TO SUCCESS

There are a lot of things the team and the innovation leader have to do to succeed. A full description of all the relevant items has already been given However, some of the subtopics are repeated here for emphasis. They are:

1. Responsive To Short Term Company Strategy
2. Short Implementation Time Frames
3. Complete Justification
4. Due Diligence
5. Sensitivity To Customer Needs
6. Credibility
7. Planning
8. Preventative Management
9. Teaming
10. Communication
11. Motivation

A detailed description of these, in addition to earlier in this paper, has been written and reported [10, 18].

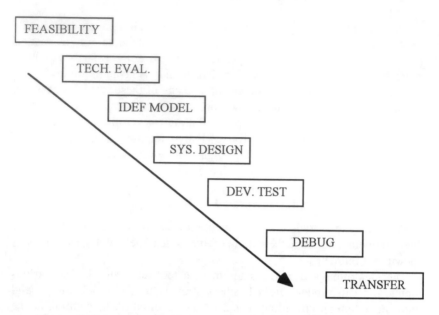

Figure 5. Concurrent Implementation of Various Stages of Systems
Development.

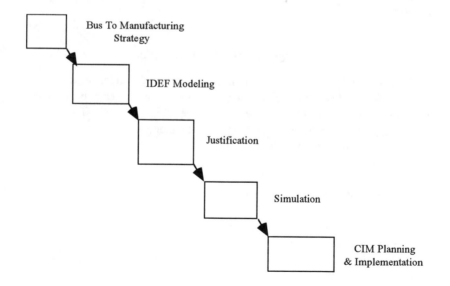

Figure 6. Integrated Methodology.

6. RESEARCH ISSUES

Important research issues of this domain are:

1. Concurrent Implementation
2. Integrated Methodology
3. Prototyping
4. Experience of the team and the leader

Each of these have been mentioned appropriately and relavantly, as above in their due context. However, they are briefly being touched upon below.

6.1. Concurrent Implementation

Life cycle of these projects can be split in 3 classical Phases. Phase 1, is the general design and due diligence phase. Phase 2 is the detailed design and specification phase. The last phase, Phase 3, is the implementation phase. The outline of each one of these 3 phases is as per Gintic Institute of Manufacturing Technology, given below.

Outline of Feasibility/ General Design/ Conceptual Design - Phase 1

1. Executive Summary
2. Table of Contents
3. Introduction
4. Existing Systems and Needs Analysis
5. Benefits Analysis
6. Proposed Architecture
7. Responsive Networks and Platforms
8. Responsive Software and Systems (Packaged and Custom, if any)
9. Estimate Cost and ROI
10. Implementation Plan and Schedule including Migration Path
11. Recommendations
12. Appendices (if any)

Outline of Detailed Design- Phase 2

1. Executive Summary
2. Table of Contents
3. Introduction
4. Revalidated needs and benefits
5. Current Operations Modeling

6. Detailed Information Flow Modeling
7. Future Operations Modeling
8. Detailed Information Flow Modeling
9. Studies in Simulations- Validation of Future Operations Models
10. (Covering Minor Development and Major Procurement)
    (Do following for every area of the enterprise)
    - Detailed Hardware and Software Specifications
    - Policy and Procedures Development
    - Input Parameter Specifications per the procured system
    - Output Parameter Specifications per the procured system
    - Algorithm and Processing Specifications per the procured system
    - Communication and Computing Interface and Standards Specifications

11. Vendor Evaluation and Selection

    - Request for Proposal Document
        * Functional Requirements
        * Performance Requirements
        * Acceptance Criteria
        * Integration Requirements
        * Interface and Standards Requirements
    - Evaluation Matrix with Weightings
    - Tenders
    - Results of bid evaluation
    - Reasons for Selection and Awarding the contract

12. Detailed Implementation Plan and Schedule
13. Cost for Next Phase
14. Recommendations
15. Appendices (if any)

Outline of Development & Implementation- Phase 3

1. Executive Summary
2. Table of Contents
3. Introduction
4. Hardware and Software specifications
5. Policy and Procedures (if condition 8B of Phase 2 outline applies)
6. Customization
7. Integration Plan and Integration Procedure

8. Test Plan and Test Procedures
9. Test Facilities
10. Acceptance Test Plan and Test Procedure
11. Technology Transfer
  A. Training Plan and Procedure(of operators/users)
  B. Operating Instructions Documentation
  C. Consultancy and Maintenance

12. Recommendations
13. Appendices
  A. Data Bases
  B. Programs
  C. GUI

While the systems community has done the 3 phases sequentially, lately customer driven pressure has demanded that some, or all, of these phase activities be done concurrently. There is 1 Gintic project, with a local SME, where Phase 2 and Phase 3 is being handled concurrently. The detailed learning will be reported [24]. Suffice it to say, however, that these are uncharted waters and further investigations can be undertaken to explore the issues and the problems and how to resolve them.

6.2. Integrated Methodology

Building CM from Modeling and Simulating the enterprise to implementing has been an art. Those who were experienced could do it. Those who were not experienced learned their lessons the hard way.

However, the need for "a model driven toolset for supporting the life cycle of CIM systems", has been well recognized [25]. These efforts tend to explore the integration tool sets, which embrace a number of different methods, created to provide computer assistance to personnel responsible for manufacturing systems during their life cycle. One such an effort is underway at Gintic. It is known as GEMs, Generic Manufacturing Systems [24]. The overview of salient goals is given below:

Overview of Gintic's GEMs
  1. Method Doc Phase 1 For Ent. Wide Manf Sys
  2. Sample Rep Phase 1 For Ent Wide Manf Sys
  3. Method Doc Phase 2 For Ent. Wide Manf Sys
  4. Sample Rep Phase 2 For Ent Wide Manf Sys
  5. Method Doc Phase 3 For Ent. Wide Manf Sys
  6. Sample Rep Phase 3 For Ent Wide Manf Sys
  7. Method Doc Phase For A Sub Sys

8. Sample Rep Phase1 For Ent Wide Manf Sys
9. Automated Life Cycle Tool Set
        Identify Contemporary Tool Sets For Evaluation
        Evaluate Cotemporary Tool Sets
        Selection Of Contemporary Tool Sets
        Piloting At A Beta Site
        Recommendations With Areas Identified If Any
        Report
10. Reference Functional Model
11. Reference Architectures
12. Reference Data Models
        Reference Info Flo Models
        Incl Triggers, Communications,
13. Sources and Triggers
14. Standards
15. Examples Of State Of The Art

6.3. Prototyping

Rapid prototyping is a well understood methodology. Tools to support the various life cycle items have already been identified. The ultimate aim of research here is to produce executable code from the requirements specification.

6.4. Experience of the Team and the Leader

In the meantime, while these tools become available, experience of the team and the leader is the only fallback option. It should be realized that replication of this experience is not easy, nor is it fast. Hence, the research in it the issues mentioned should be prioritized and undertaken. One important aspect of academic research is the lack of concentration on communication and motivation. For both the team as well as the leader these have to be emphasized. Yassin Sankar's [13] book is outstanding in dealing with the subject matter.

7. CONCLUSIONS

In the changing paradigms of manufacturing and corporate strategies, Managing Innovation while justifying and implementing large scale Mega CIM and Plant Wide Control Systems, has been discussed. What are they looking for in supporting major innovations, what are the hurdles to deal with and what are the strategies to be adopted, has been articulated in the justification and implementation stages, both from the management and technologist's point of view. Finally, the research issues have been mentioned and articulated where they arose in the context. Hope the message was clear: while not easy, in fact very difficult due to a large number of structural reasons, Mega CIM projects can be justified and can be successfully implemented.

REFERENCES

1. **Curtain F.T.**; New costing methods needed for manufacturing technology; Management Reviews 73, 29- 30; 1984.
2. **Curtain F.T.**; Planning and justifying factory automation systems; Prod Eng 31, 46- 51; 1984.
3. **Evans D.** and **Schwab P.C.**; Integrated manufacturing financing- the backing to go forward; Prod Eng, 122- 124; September 1984.
4. **Michael G.J., Millen Robert A**. Economic justification of modern computer based factory automation equipment- a status report; Proc First ORSA/TIMS Special Interest Conf on FMS; Ann Arbor, Michigan, 30- 35; 1984
5. **Bennett R.G.**; What are companies spending on CIM and how are they justifying these expenditures; Proc CIMCON '85, Dearborn, Michigan; April 1985.
6. **Canada J.R.**; Non traditional method for evaluating CIM opportunities. Assigns weights to intangibles; Ind Eng, 66- 71; March 1986.
7. **Meredith J.R.** and **Suresh N.C.**; Justifying techniques for advanced manufacturing technologies; Int J Prod Res 25, 1043- 1057; 1986.
8. **Kaplan R.S.**; Accounting the Obsolescence of cost accounting systems; California Management Review, 27; 178- 199; 1986.
9. **Hays, Wheelwright, Clark**; Dynamic Manufacturing- Creating the Learning Organization; Free Press; 1988.
10. **Bansal S.**; ALCOA's CIM Showcase, Advanced Manufacturing Systems (AMS 86), Chicago; April 1988.
11. **Bansal S.**; Framework for Plant Wide Process Control Investment Strategies, Control, Chicago; January 1991.
12. **Bansal S.**; Developing a Framework for Plant Wide Control Investment Strategy, Invited Paper for Computer Integrated Manufacturing Systems, special issue on Modelling and Justifying Technological Innovation Processing in Manufacturing, UK; November 1991.
13. **Yassin Sankar**; Management of Technological Change; John Wiley & Sons Inc; 1991.
14. **Hamid R. Parsei** and **Anil Mital**; Economics of Advanced Manufacturing Systems; Chapman and Hall; 1992
15. **Roy L. Harmon**; Reinventing the Factory- Managing the World Class Factory; Free Press; 1992

16. **Bansal S.**; Managing Innovations in CIM Environments, Invited paper at CIRP/IFORS, Capri, Italy; September 1993.
17. **Bansal S.**; A Framework for Justification and Implementation, Invited Tutorial at International Conference of Computer Integrated Manufacturing, Singapore; September 1993.
18. **Bansal S.**; Computer Integrated Manufacturing- A Greenfield Paradigm Implementation, Industrial Automation Journal, Singapore; October 1993.
19. **Michael Hammer**, James Champy; Reengineering the Corporation; Harper Collins; 1993.
20. **Bansal S.**; Management of Plant wide Automation, Industrial Automation 94, Singapore; March 1994.
21. **Gregory A. Hanson**; Automating Business Process Reengineering; Prentice Hall; 1994.
22. **Dorine C. Andrews**, **Susan K. Stalick**; Business Reengineering, the survival guide; Yourdon Press; 1994.
23. **Bansal S.**; Practical Applications of Models and their Benefits , Evolution of a New Paradigm, Invited Paper for Journal of International Federation of Automatic Controls (CEP); October1994.
24. **Bansal S.**; Gintic Methodology of supporting Life Cycle of Manufacturing Information Systems; To be Published.
25. **Edwards J.M.**, **Murgatroyd S.** and **Weston R.H.**; A Model Driven Toolset for Supporting the Life Cycle of CIMSystems; To be Published; ICCIM; Singapore; 1995.

Index